Titles in This Series

13 **V. P. Maslov and S. N. Samborskiĭ, editors,** Idempotent analysis, 1992
12 **R. Z. Khasminskiĭ, editor,** Topics in nonparametric estimation, 1992
11 **B. Ya. Levin, editor,** Entire and subharmonic functions, 1992
10 **A. V. Babin and M. I. Vishik, editors,** Properties of global attractors of partial differential equations, 1992
9 **A. M. Vershik, editor,** Representation theory and dynamical systems, 1992
8 **E. B. Vinberg, editor,** Lie groups, their discrete subgroups, and invariant theory, 1992
7 **M. Sh. Birman, editor,** Estimates and asymptotics for discrete spectra of integral and differential equations, 1991
6 **A. T. Fomenko, editor,** Topological classification of integrable systems, 1991
5 **R. A. Minlos, editor,** Many-particle Hamiltonians: spectra and scattering, 1991
4 **A. A. Suslin, editor,** Algebraic K-theory, 1991
3 **Ya. G. Sinaĭ, editor,** Dynamical systems and statistical mechanics, 1991
2 **A. A. Kirillov, editor,** Topics in representation theory, 1991
1 **V. I. Arnold, editor,** Theory of singularities and its applications, 1990

Advances in
SOVIET MATHEMATICS

Volume 13

Idempotent Analysis

V. P. Maslov
S. N. Samborskiĭ
Editors

American Mathematical Society
Providence, Rhode Island

Translation edited by A. B. SOSSINSKY

1991 *Mathematics Subject Classification.* Primary 16Y20, 16Y60, 20M20, 35B40, 35F20, 47B38, 49L05, 49L20, 49L99, 90C27; Secondary 16Y60, 49B99, 49N60, 81Q20, 90A16, 90C27.

Library of Congress Catalog Card Number: 91-640741

ISBN 0-8218-4114-9; ISSN 1051-8037

The paper used in this book is acid-free and falls within the guidelines
established to ensure permanence and durability. ♾
This publication was typeset using AMS-TEX,
the American Mathematical Society's TEX macro system.
10 9 8 7 6 5 4 3 2 1 97 96 95 94 93 92

Contents

Idempotent Analysis
(In Place of an Introduction)

Idempotent analysis is a new branch of mathematical analysis which studies functional spaces and their mappings when the algebraic structure is generated by an idempotent operation (the operation denoted by, say, $\oplus$ is idempotent if $a \oplus a = a$ for all a). In classical analysis the main algebraic structure, the basic foundation that supports the stage on which the action is played out, is the structure of a field ($\mathbb{R}$ or $\mathbb{C}$). Let us delete the requirement of the existence of inverse elements for addition from the list of axioms for fields. The semiring structure thus obtained is too general to serve as the basis of the analysis for functions with values in this semiring. However, if the deleted property of "addition" is replaced by the idempotence requirement, the structure acquires sufficient rigidity to advance in analysis and even to advance surprisingly far: in the case of "linear" (in the sense of the new operations) maps, the analogs of many basic facts of functional analysis turn out to be valid (and nontrivial), facts such as the Riesz and Hahn-Banach theorems, Fourier transforms, spectral analysis, the operational calculus, up to the theory of distributions.

These analogs not only provide a convenient background for our "geometric intuition", but their understanding often suggests nontrivial constructions, which could hardly have been invented if we did not "stand on the shoulders of the giants" of classical analysis.

The possibility of obtaining a meaningful and aesthetically attractive theory here combines with an abundant source of examples of mappings whose "linearity" occurs only in the idempotent structures. Since these mappings are often not even smooth and are poorly approximated by locally linear maps, they cannot be successfully studied by the methods of traditional analysis, but their study cannot be avoided when we consider optimization problems, convex analysis, game theory, mathematical economics, etc.

Semirings with idempotent operations have been considered a good many years ago by many authors, especially in connection with complexity estimates of computational algorithms in optimization problems. The fact that many optimization problems are "linear" in the appropriate semirings made possible the use of the computational techniques of linear algebra for their solution. We cannot present here a review of the relevant literature, since it is dispersed in many publications, often appears in implicit form, with different terminology in each case. We therefore limit ourselves to mentioning the monograph by Aho, Hopcroft, and Ullman [1] and the survey by Gondran [2].

However, the articles of the present collection are not related to the continuation and development of this line of study. They are united by their involvement with the properties of a mathematical object that has made its appearance only recently. We have in mind the analysis of "linear operators" in function spaces with values in an idempotent semiring. (Apparently, the exposition of this approach and its motivation first appeared in V. P. Maslov's monograph [3] and his article [4].)

Rather than summarizing the contents of the articles in this collection, we shall mention here some of the critical ideas and observations around which, to a certain extent, the results presented here are concentrated.

Let the idempotent semiring (replacing the base field in ordinary calculus) be the numerical semiring

$$\mathfrak{R} = \mathbb{R} \cup \{\infty\}$$

with operations $a \oplus b = \min(a, b)$, $a \odot b = a + b$ and neutral elements $\mathbf{0} = \infty$, $\mathbf{1} = 0$ respectively. Then, in the space of functions defined on some set X and ranging over $\mathfrak{R}$, mappings of the form

$$(Af)(x) = \inf_{y \in X}(k(x, y) + f(y)) \stackrel{\text{def}}{=} \bigoplus \int_X k(x, y) \odot f(y)\, dy \tag{1}$$

are "linear" (i.e., $A(\lambda \odot f \oplus g) = \lambda \odot Af \oplus Ag$); moreover, (1) is the exact analog of a linear integral operator with kernel k, and this statement can be given exact meaning in the framework of the theory of idempotent integration (here and below quotation marks are used to indicate notions well known in classical analysis, but in whose definition the ordinary operations $+$ and $\cdot$ have been replaced by the new operations $\oplus$ and $\odot$ from the corresponding semiring). Maps of the form (1) naturally arise in optimization problems; they occur, say, in Bellman's recursive relations: if $\varphi_n(x)$ is the minimal value of the optimization criterion in the control problem of dynamics on n-step trajectories originating at a fixed point of the phase space X, then $\varphi_n = A^n \varphi_0 = A\varphi_{n-1}$.

The multicriterial problems of control dynamics similarly lead to operators "linear" in more complicated idempotent semirings, e.g. in the semirings of Pareto sets [5].

Consider the Cauchy problem for the Hamilton-Jacobi equation,

$$\frac{\partial y}{\partial t} + H(x, Dy) = 0, \qquad y(0, x) = y_0(x), \tag{2}$$

where y is a real-valued function of $t \in \mathbb{R}$, and x comes from a domain $X \subset \mathbb{R}^n$, while $D = (\partial/\partial x_1, \dots, \partial/\partial x_n)$. For well-known reasons, the solution of the problem (2) should be looked for in the class of functions of a certain generalized smoothness (which we call, putting aside considerations of rigor, piecewise smoothness). Denote by U_t the map that assigns to each given $y_0(\cdot)$ the solution $y(t, \cdot)$ of problem (2) at the moment of time t.

The fundamental circumstance determining the properties of the solutions of problem (2) is the fact that the map U_t for each t is a "linear integral operator" in the "linear space" of functions on X with values in the numerical semiring $\mathfrak{R}$. This circumstance allows us to use, in the definition of "distributions", the "adjoint operator" with respect to the following scalar product:

$$(f, g) = \bigoplus \int_X f(x) \odot g(x)\, dx = \inf_{x \in X}(f(x) + g(x))$$

and use "weak convergence":

$$f_i \to f \quad \text{iff} \quad (f_i, \varphi) \to (f, \varphi) \text{ in } \mathfrak{R}$$

for all φ from a certain space of "test functions" possessing ordinary smoothness [6]. (Let us note also that "weak convergence" gives mathematical interpretation to the idea of modulated signal in radiophysics.) Indeed, the "weak limit" of a rapidly oscillating function is its lower (in the sense of the semiring $\mathfrak{R}$) evolute [3, 4].

Now suppose that y is an "eigenfunction" of the "linear operator" U_t corresponding to the "eigenvalue" λ, i.e., a function y such that $U_t y = \lambda \odot y$. It can be shown that the "linear operators" U_t for different t have "eigenvalues" of the form $\lambda_0 t$, where λ_0 does not depend on t, and the same "eigensubspace" E. Let $\bar{y}$ be a function from E. Then $(t, x) \to \lambda_0 t + \bar{y}(x)$ is a solution of problem (2) with initial data $\bar{y}$, and therefore, $\bar{y}$ is a solution in the sense of "distributions" of the equation

$$\lambda_0 + H(x, Dy) = 0. \tag{3}$$

All this occurs, of course, only because the "distributions" which satisfy the (nonlinear!) equation (3) constitute a "linear subspace". The same fundamental circumstance is the basis of a construction that repeats essentially the construction of Sobolev spaces, but with the algebraic structure of the idempotent semiring $\mathfrak{R}$ replacing that of fields $\mathbb{R}$ or $\mathbb{C}$. This construction results in spaces of "generalized differentiable functions" that are semimodules over $\mathfrak{R}$. Topological considerations single out in these semimodules certain subsemimodules of functions where existence and uniqueness theorems for solutions of boundary value problems for Hamilton-Jacobi and Bellman equations hold. (Let us remark that in C^1 we do not have the existence of solutions of these problems, and in the Sobolev space W^1 we do not have the uniqueness.)

We have already mentioned "eigenfunctions". Their existence follows from fairly general facts of "special theory" [7, 8]. For compact "integral operators", the "spectrum" is nonempty and consists of a finite or countable set of isolated points whose only possible limit point is $\mathbf{0}$. Except for the possibility of infinite multiplicity, this fact closely resembles the Riesz-Schauder theorem of classical functional analysis. "Spectral" terminology, as in classical analysis, turns out to be very convenient for asymptotic estimates of iterations of "linear operators" [7].

Finally, let us consider "Fourier transforms". Let $\operatorname{Char} G$ be the set of "characters" of a group G, i.e. "multiplicative functionals" from G into a semiring (in the case of the numerical semiring $\mathfrak{R}$, G being an additive subgroup of $\mathbb{R}^n$, this is the set of linear functionals). Then let us define the "Fourier transform" of f on G by the "usual" formula

$$(F, g)(\chi) = \bigoplus \int_G \chi(g) \odot f(g)\, dg = \inf_{g \in G} (\chi(g) \odot f(g)).$$

This transformation takes the "convolution" of functions to pointwise "multiplication" and possesses the analogs of many important properties of ordinary Fourier transformations. In the case of the numerical semiring $\mathfrak{R}$ and $G \subset \mathbb{R}^n$, the "Fourier transform" coincides with the classical Legendre transform (see [4], although this observation has been part of the folklore for a sufficiently long time). For the case of semirings of Pareto sets, the "Fourier transformation" reduces the multicriterial optimization problem to a family of singlecriterial ones [5].

These observations, as well as many others, allow us to claim that idempotent analysis is called upon to play the same unifying role in many branches of mathematics related to external phenomena and structures that in its day was played by functional analysis for mathematical physics, numerical methods, and differential equations. To carry out this unification, it is necessary to study the algebraic and analytic structures appearing in function spaces (and in their spaces of maps) consisting of functions with values in idempotent semirings.

We hope that the articles in this collection constitute an advance in this direction.

References

1. A. Aho, J. E. Hopcroft, and J. E. Ullman, *The design and analysis of computer algorithms*, Addison-Wesley, Reading, Ma., 1976.
2. M. Gondran, *Path algebra and algorithms*, Combinatorial Programming: Methods and Applications (B. Roy, ed.), NATO Adv. Study Inst. Ser., Ser. C: Math. and Phys. Sci., vol. 19, Reidel, Dordrecht, 1975, pp. 137–148.
3. V. P. Maslov, *Asymptotic methods for solving pseudodifferential equations*, "Nauka", Moscow, 1987. (Russian)
4. ———, *On a new superposition principle for optimization problems*, Uspekhi Mat. Nauk **42** (1987), no. 3, 39–48; English transl. in Russian Math. Surveys **42** (1987).
5. S. N. Samborskiĭ and A. A. Tarashchan, *On semirings arising in multicriterion optimization problems and problems of analysis of computational media*, Dokl. Akad. Nauk SSSR **308** (1989), no. 6; English transl., Soviet Math. Dokl. **40** (1990), no. 2, 441–445.

6. V. N. Kolokoltsov and V. P. Maslov, *Idempotent calculus as the apparatus of optimization theory,* I, Funktsional. Anal. i Prilozhen. **23** (1989), no. 1, 1–14; II, **23**, no. 4, 53–62; English transl. in Functional Anal. Appl. **23** (1989).
7. P. I. Dudnikov and S. N. Samborskiĭ, *Spectra of endomorphisms of semimodules over semirings with an idempotent operation,* Dokl. Akad. Nauk SSSR **308** (1989), no. 4, 363–366; English transl. in Soviet Math. Dokl. **40** (1990).
8. ———, *Endomorphisms of semimodules over semirings with an idempotent operation,* Izv. Akad. Nauk SSSR Ser. Mat. **55** (1991), 93–109; English transl. in Math. USSR-Izv. **38** (1992).

V. P. Maslov and S. N. Samborskiĭ

Translated by A. B. SOSSINSKY

ADVANCES IN SOVIET MATHEMATICS
Volume 13, 1992

Quantization of the Bellman Equation, Exponential Asymptotics and Tunneling

S. YU. DOBROKHOTOV, V. N. KOLOKOLTSOV, AND V. P. MASLOV

§1. Introduction

The famous WKB method allows one to construct local asymptotic solutions of linear pseudodifferential equations

$$ih\frac{\partial \psi}{\partial t} = H\left(x, \frac{h}{i}\frac{\partial}{\partial x}\right)\psi \tag{1.1}$$

by means of smooth solutions of the characteristic first order equation:

$$\frac{\partial S}{\partial t} + H\left(x, \frac{\partial S}{\partial x}\right) = 0. \tag{1.2}$$

In particular, one must solve the equation

$$\frac{\partial S}{\partial t} + \frac{1}{2}\left(\frac{\partial S}{\partial x}\right)^2 + V(x) = 0 \tag{1.3}$$

to construct asymptotic solutions of the Schrödinger equation

$$ih\frac{\partial \psi}{\partial t} = \left(-\frac{h^2}{2}\Delta + V(x)\right)\psi. \tag{1.4}$$

When constructing the global asymptotics of equation (1.1), one must use all the branches of the multivalued solutions of equation (1.2). These multivalued solutions can be represented geometrically by means of the so-called Lagrange manifolds. Thus the successive generalization of the WKB method leads one to the quantization of Lagrange manifolds and to the construction of the canonical operator that gives the global asymptotic solution of equation (1.1) [3–7]. But to construct the asymptotics for the low lying eigenfunctions of the Schrödinger operator another approach is more useful. Instead of equation (1.3), let us consider the equation

$$\frac{\partial S}{\partial t} + \frac{1}{2}\left(\frac{\partial S}{\partial x}\right)^2 - V(x) = 0 \tag{1.5}$$

1991 *Mathematics Subject Classification.* Primary 35B40, 35F20, 90C27; Secondary 81Q20.

1051-8037/92 \$1.00 + \$.25 per page

with inverted potential and let us search for its generalized (not smooth everywhere) but single-valued solution, choosing the minimal value among all branches of the multivalued smooth solution. These generalized solutions (which are at the same time the solutions of some optimization problem) can be well defined by means of idempotent analysis [9] or by means of the well-known viscosity method. It turns out that the solutions of the Cauchy problem for equation (1.5) determine the quasiclassical phase (or, more precisely, the entropy) of asymptotic solutions of the equation

$$h\frac{\partial u}{\partial t} = \left(\frac{h^2}{2}\Delta - V(x)\right)u \tag{1.6}$$

that can be obtained from (1.4) by passing to imaginary time and the stationary solutions of (1.5) (or, in a more general situation, the eigenfunctions of the resolving operator of equation (1.5) in the sense of idempotent calculus) and determine the phases of the low eigenfunctions of the Schrödinger operator. The corresponding geometric quantization methods leads to the construction of the tunnel canonical operator. This construction first appeared in [1, 2] and was developed in [25]. Speaking more generally, the tunnel canonical operator method allows one to construct exponential asymptotic solutions of the equations

$$h\frac{\partial u}{\partial t} = H\left(t, x, -h\frac{\partial}{\partial x}\right)u, \tag{1.7}$$

where

$$H\left(t, x, -h\frac{\partial}{\partial x}\right)u = (2\pi h)^{-n}\int_{\mathbb{R}^{2n}} \exp\left\{\frac{i}{h}\langle p, x-\xi\rangle\right\}H(t, x, -ip)u(t, \xi)\,d\xi\,dp \tag{1.8}$$

for a certain class of Hamiltonians H. This class was first introduced in [3] and was called there the class of tunnel Hamiltonians. The precise definition appeared in [1, 2]: equation (1.7) is called an equation of *tunnel* type, if

i) $H(t, x, p)$ is a smooth function in $x \in \mathbb{R}^n$, $t \in \mathbb{R}_+$ and an analytic function in $p \in \mathbb{C}^n$,
ii) the Lagrange function $L = \langle p, \partial H/\partial p\rangle - H$ is nonnegative,
iii) the matrix of second derivatives $\partial^2 H/\partial p^2$ is nondegenerate for all real p,
iv) $H(t, x, p) = \max_{\eta\in\mathbb{R}^n} \operatorname{Re} H(t, x, p+i\eta)$ for all $x, p \in \mathbb{R}^n$.

Equation (1.6) with nonnegative V gives the simplest example of the tunnel type equation. Another example is the Kolmogorov-Feller equation

$$\frac{\partial u}{\partial t} = b(t, x)\frac{\partial u}{\partial x} + \frac{1}{h}\int_{\mathbb{R}^n}(u(t, x+h\xi) - u(t, x))\,d\mu_{t,x}(\xi),$$

where $d\mu_{t,x}$ is a probability measure on $\mathbb{R}^n$ for all t, x, and $b(t, x)$ is a given vector-function. The notion of tunnel equation can be generalized to the case of systems: a system of the form (1.7) with a matrix-valued H is called a *tunnel system* if its eigenvalues are functions of the tunnel type. An important example is the linearized Navier-Stokes system with small viscosity

$$\frac{\partial v}{\partial t} + \langle u, \nabla\rangle v + a^2\nabla\rho - c_1 h\nabla^2 v - c_2 h\nabla \operatorname{div} v = 0,$$
$$\frac{\partial \rho}{\partial t} + \operatorname{div} v + \langle u, \nabla\rho\rangle = 0,$$

where $v = (v_1, v_2, v_3)(t, x)$, $\rho(t, x)$ are unknown functions and $u = (u_1, u_2, u_3)(t, x)$, $a(t, x)$ are given smooth functions, c_1, c_2 being positive physical constants. For other examples, see [1, 2]. The solutions of the Cauchy problem for tunnel equations with finite initial data are exponentially small in h almost everywhere in the phase space. This is their main property. Let us point out, however, that the tunnel properties i)–iv) are certainly not necessary for that. For instance, the equation

$$h\frac{\partial u}{\partial t} = -\frac{h^4}{4}\frac{\partial^4 u}{\partial x^4} - u \tag{1.9}$$

with $x \in \mathbb{R}$ is not of tunnel type, but its Green function $G(t, x-\xi)$ has the form $G(t, x-\xi) = f(t, |x-\xi|)$ for $x \neq \xi$, $t \neq 0$, where

$$f(t, y) = (6\pi h)^{-1/2} y^{-1/3} t^{-1/6} \exp\left\{-\frac{1}{h}\left(t + \frac{3}{8}y\left(\frac{y}{t}\right)^{1/3}\right)\right\}$$
$$\times \sin\left\{\frac{1}{h}\left(1 + \frac{3\sqrt{3}}{8}y\left(\frac{y}{t}\right)^{1/3}\right)\right\}(1 + O(h)),$$

and thus G is exponentially small in h almost everywhere. An interesting problem is to find general conditions for Hamiltonians for which the Green function of equation (1.7) is exponentially small in h almost everywhere.

The exponential asymptotics of equation (1.6) give the asymptotic solution of the large deviation problem from probability theory. These asymptotics were first obtained in [12, 17], but under the assumption of absence of focal points, and then in [1, 2] in the general case.

Our paper is devoted to the derivation of the asymptotics of the low eigenfunctions of the Schrödinger operator obtained by investigating large time asymptotics for equation (1.6). The relationship between these problems is the consequence of the following elementary argument. Let us consider equation (1.6). Let $V(x)$ be a nonnegative function that increases as $|x| \to \infty$. Then the spectrum of the operator $-h^2\Delta/2 + V(x)$ in $L_2(\mathbb{R}^n)$ is discrete, and if we denote by $\{\psi_k\}$ and $\{E_k\}$ the corresponding orthonormal functions and values, then the solution of the Cauchy problem (1.6) with initial function $u|_{t=0} = u_0(x)$ has the form

$$u = \sum_{k=0}^{\infty} c_k e^{-E_k t/h}\psi_k, \tag{1.10}$$

where

$$c_k = (\psi_k, u_0) = \int_{\mathbb{R}^n} \psi_k u_0 \, dx.$$

If $c_0 \neq 0$, we may divide (1.10) by $c_0 \exp E_0 t/h$ and get

$$\psi_0 = \lim_{t\to\infty} \frac{ue^{E_0 t/h}}{(u_0, \psi_0)}. \tag{1.11}$$

The relationship between the Cauchy problem for the nonstationary equation (1.6) and the spectral problem

$$-(h^2/2)\Delta\psi + V(x)\psi = E\psi \tag{1.12}$$

is based on formula (1.11). It is clear that the exact solution of the Cauchy problem for equation (1.6) and the exact function (1.11) can be found only in exceptional cases, and for arbitrary $V(x)$ analytic formulas can only be asymptotic. If $V(x) \geqslant 0$, the solution $u(t, x, \xi)$ of the Cauchy problem (1.6) with initial function $u|_{t=0} = \delta(x - \xi)$, i.e., the fundamental solution, and the eigenfunctions of (1.12) for $E(h) \to 0$ as $h \to 0$ are proved to be exponentially small in $h \to 0$ almost everywhere and almost everywhere are given by the formulas

$$u = h^{-n/2} e^{-S(t, x, \xi)/h} (\varphi(t, x, \xi) + O(h)), \tag{1.13}$$

$$\psi = h^{-n/2} e^{-S(x)/h} (\varphi(x) + O(h)), \tag{1.14}$$

where S, φ are continuous functions, $S \geqslant 0$.

One of the major results of this article consists in the proof of the fact that the asymptotics constructed hold not only for the "standard" (in asymptotic theory) values of time $t = O(1)$ (as $h \to 0$), but also for "very large" values of time $t = h^{-(1+\varkappa)}$, $\varkappa \in (0, 1)$. It is precisely this result that allows us to pass rigorously from the asymptotics of the fundamental solution of equation (1.6) to the asymptotics of the eigenfunctions of the Schrödinger operator which detect tunnel effects.

Let us explain this more explicitly. We suppose that ξ is a nondegenerate minimum point of the potential $V(x)$ and $V(\xi) = 0$. Since the potential V in a neighborhood of ξ can be approximated by the harmonic oscillator potential

$$V_{\text{osc}} = \left(x - \xi, \frac{\partial^2 V(\xi)}{\partial x^2}(x - \xi)\right),$$

we can find a series of asymptotic eigenfunctions and eigenvalues of the initial operator $-h^2\Delta/2 + V(x)$, their leading term coinciding with the eigenfunctions and the eigenvalues of the quantum harmonic oscillator. In particular, the lowest asymptotic eigenvalue and eigenfunction have the form

$$E_0 = \frac{h}{2}\sum_{j=1}^{n} \omega_j, \qquad \psi_0 = (\pi h)^{-n/4}(\omega_1 \cdots \omega_n)^{-1/2} \exp\{-S(x)/h\}, \tag{1.15}$$

where

$$S = \frac{1}{2}\left(x - \xi, \left(\frac{\partial^2 V(\xi)}{\partial x^2}\right)^{1/2} (x - \xi)\right)$$

and ω_j^2 are the eigenvalues of the matrix $\partial^2 V(\xi)/\partial x^2$, $\omega_1, \dots, \omega_n > 0$.

The asymptotic formulas (1.15) have been used with success in the physical literature for a long time. In [18] (or in [25]) one can find the rigorous arguments which show the relationship between the eigenfunctions and the eigenvalues of the quantum harmonic oscillator and of the operator $-h^2\Delta/2 + V(x)$. Note that (1.15) gives only power asymptotics in h of the real eigenfunction of the operator $-h^2\Delta/2 + V(x)$ and is entirely meaningless outside a neighborhood of the minimum point ξ. Therefore it is useless when investigating tunnel effects.

Let $V(x)$ be symmetric (either with respect to the point $x = 0$, or with respect to some hyperspace) and have two points $\xi_{+,-}$ of global minimum. Then near the minimal eigenvalue E_0 of the operator $-(h^2/2)\Delta + V(x)$ there is another eigenvalue E_1 that differs from E_0 by a value exponentially small in h. Precisely, the following formulas hold:

$$E_{0,1} = \frac{h}{2}\sum_{j=1}^{n} + O(h^2),$$

$$E_1 - E_0 = A(h)\exp\left\{-\frac{1}{h}\int_{\xi_-}^{\xi_+} p\,dx\right\}, \tag{1.16}$$

where $p(t) = \dot{X}$ and $x = X(t)$ is the trajectory of the Newton system

$$\ddot{x} = \partial V/\partial x \tag{1.17}$$

joining the points ξ_- and ξ_+: $X(\mp\infty) = \xi_\mp$. (If there are several such trajectories, one must take the sum of the corresponding terms for each of these trajectories in the right-hand side of (1.8).) The value $A(h)$ is also defined via the solutions $X(t)$ and the solutions of the corresponding system in variations. The trajectory $X(t)$ is called an *instanton*. It seems that formula (1.8) first appeared in the book [5] and papers [21, 29] and was actively used afterwards in quantum field theory, most of the papers dealing either with the one-dimensional ($x \in \mathbb{R}$) [26–27, 29–31, 33] or with infinite-dimensional case (for a survey, see [37] or [23]). A mathematical proof in the n-dimensional case appeared in [2, 19]. Some generalizations were given in [32].

In this paper we give, in particular, a complete exposition of the method that was pointed out in [2] and thus obtain a formula for the coefficient $A(h)$ in (1.16) (which is lacking in [19]). Our argumentation is quite different from the methods of [23], based on the use of the Feynman integral and the Agmon metric. The present paper is an improved and completed version of our article [25].

Let us describe the content of the paper in a more detailed way. In §2 we construct the asymptotics of the fundamental solution of the parabolic equation (1.6) in the small. We try to find the solution of (1.6) in the form

$$u = (2\pi h)^{-n/2} e^{-S(t,x)/h} (\varphi_0(t, x) + h\varphi_1(t, x) + \cdots).$$

The standard WKB-procedure gives the Hamilton-Jacobi equation (1.3) and the transport equation

$$\frac{\partial \varphi_0}{\partial t} + \frac{\partial S}{\partial x}\frac{\partial \varphi_0}{\partial x} + \frac{1}{2}\varphi_0 \Delta S = 0.$$

The initial conditions for S and φ_0 are chosen so that $u \to \delta(x - \xi)$ as $t \to 0$. The corresponding solutions S and φ are expressed in terms of the solution of a variational problem and of the Newton system (1.17). We then get an important estimate for the difference between the exact solution u and the asymptotics u_{as}:

$$|u - u_{\mathrm{as}}| = e^{-S/h}\varphi_0 O(ht^3). \tag{1.18}$$

The global asymptotics of the fundamental solution of equation (1.6) is given by the tunnel canonical operator defined in §3. We develop there a geometric approach to the construction of asymptotics (Lagrange manifolds). The main obstacle to the construction of the asymptotics "in the large" is well known. It is connected with the appearance of focal points and of domains where the function S is multivalued. The specific character of "tunnel" problems, in particular, consists in the fact that in contrast to "ordinary" quasiclassical asymptotics, in "tunnel" problems only one term brings a contribution to the result, namely the term that corresponds to the branch of S with minimal value at the given point x. The global construction of the asymptotics of the fundamental solution of equation (1.6) is given in §4. Section 5 is devoted to one of the main results of the paper, that is to the proof of the fact that the asymptotics constructed above remain valid for "very large" values of time $t = h^{-(1+\varkappa)}$, $\varkappa > 0$. This proof is based on multiple use of the Laplace method (with special subtle estimates, given below) and exploits (1.18) substantially. There we also give the asymptotics (for "very large" values of time) of the special Cauchy problems for (1.6) that is necessary for further investigations of the spectral problem (1.12).

The transition from the solution of the nonstationary Cauchy problem (1.6) to the solutions of (1.12) is based, as it was already pointed out, on formula (1.11), where we put $t = h^{-(1+\varkappa)}$ instead of infinite time. This is enough to be able to delete from the series in eigenfunctions all the terms with eigenvalues outside an $O(h)$-neighborhood of the harmonic oscillator's lower energy level $E_0 = (h/2)\sum_{j=1}^{n} \omega_j$ and then to pass to the limit $t \to 0$ in the main term. This procedure takes into consideration the existence of two small parameters h and $1/t$ in the problem and allows us to define the exponential asymptotics of the low eigenfunctions of the Schrödinger operator. It implies formula (1.16) and analogous formulas for the case of several minimum points of potential.

In §7 analogous results are obtained for the case when the configuration space is a torus, in particular, a circle. In §§8 and 9 we consider some interesting examples.

Now we shall give the special estimates for the Laplace method which were used in the proof of the main results. A rather tedious proof can be found in [25]. We need an approximate formula for the integral

$$I(h) = \int_\Omega f(x)\exp\{-S(x)/h\}\,dx\,, \tag{1.19}$$

where Ω is a domain in $\mathbb{R}^n$, h is a positive real parameter, the functions f, S are real and continuous in Ω, f is nonnegative. Let us suppose that

i) the function S has only one minimum point x_0 in Ω, in some compact neighborhood U of x_0 the function S has continuous derivative of the fourth order and the matrix $S''(x_0)$ is positive;
ii) the function f has continuous derivatives of the second order;
iii) the integral $I(h)$ is defined for some $h = h_0$.

Let d and D be the minimal eigenvalue and the determinant of the matrix $S''(x_0)$ respectively and let F, F_1, F_2, C_3, C_4 be the maximum of norms f, f', f'', $S^{(3)}$, $S^{(4)}$ respectively. Let δ be a positive real number.

PROPOSITION. *There exists an $h_1 \leqslant h_0$ such that*

$$I(h) = (2\pi h)^{n/2}D^{-1/2}(f(x_0) + \alpha(h)h)\exp\{-S(x_0)/h\} \tag{1.20}$$

for $h \leqslant h_0$, where

$$\begin{aligned}|\alpha(h)| \leqslant{}& (1/2)F_2d^{-1}n + (1/4)e^{1/2}FC_3^2d^{-3/2}n(n+1)(n+2)\\ &+ ((1/4!)C_4f(x_0) + (1/3!)C_3F_1)d^{-2}n(n+2)\\ &+ f(x_0)(2\pi h_0)^{n/2}h^\delta + e^{1/3}D^{1/2}h^\delta\exp\{S(x_0)/h\}I(h_0).\end{aligned} \tag{1.21}$$

§2. Small values of time asymptotics for the fundamental solution of the Cauchy problem for the heat conduction equation with potential

We consider the equation

$$h\frac{\partial u}{\partial t} = \frac{h^2}{2}\Delta u - V(x)u\,, \qquad x\in\mathbb{R}^n\,,\ t>0\,, \tag{2.1}$$

where h is a positive parameter, Δ is the Laplace operator, the potential V is a smooth nonnegative function with bounded matrix of second derivatives: $\|V''(x)\| \leqslant C$ for all $x\in\mathbb{R}^n$ and a fixed C.

We shall construct the asymptotics as $h\to 0$ of the fundamental solution of the Cauchy problem for this equation, i.e., the solution $u(t,x) = u(t,x,\xi,h)$ which satisfies the initial condition

$$u(t,x,\xi,h) = \delta(x-\xi). \tag{2.2}$$

At first we shall find this asymptotics for small t. For this purpose we need some estimates and identities for the solutions of the system in variations for the Newton equation.

Take the Hamiltonian $H(x, p) = p^2/2 - V(x)$ and denote by $X(t, x, p)$, $P(t, x, p)$ the solution of the Hamiltonian system

$$\dot{x} = p, \qquad \dot{p} = \frac{\partial V}{\partial x} \tag{2.3}$$

with initial conditions $X(0, x, p) = x$, $P(0, x, p) = p$. Due to the properties of V, these solutions exist and are unique for all t.

We need the following well-known fact [6]. There exists a t_0 such that $\det(\partial X/\partial p) \neq 0$ for all $t \leqslant t_0$, and for all $x, \xi \in \mathbb{R}^n$ there exists a unique solution of the Hamiltonian system (2.3) with boundary conditions $X(0) = \xi$ and $X(t) = x$. On this curve the functional

$$\int_0^t (\dot{y}^2(\tau)/2 + V(y(\tau)))\, d\tau, \tag{2.4}$$

defined for all continuous piecewise smooth curves $y(\tau)$ with fixed endpoints $y(0) = \xi$, $y(t) = x$, takes its minimal value. This value will be denoted by $S(t, x, \xi)$. We also denote by $p_0(t, x, \xi)$ the momentum for which $X(t, \xi, p_0(t, x, \xi)) = x$ and denote the Jacobian by

$$J(t, x, \xi) = \det \frac{\partial X}{\partial p}(t, \xi, p_0(t, x, \xi)).$$

The main result of this section is the following.

THEOREM 1. a) *If $V(x)$ has third and fourth order derivatives bounded in $\mathbb{R}^n$, then the solution of the problem* (2.1), (2.2) *for $t \in (0, t_0]$ is*

$$u(t, x, \xi, h) = (2\pi h)^{-n/2} J(t, x, \xi)^{-1/2} \exp\{-S(t, x, \xi)/2\}(1 + O(ht^3)), \tag{2.5}$$

where $O(ht^3)$ is uniform for all $x, \xi \in \mathbb{R}^n$.

b) *If V has derivatives of higher orders, then the function $O(ht^3)$ in* (2.5) *remains $O(ht^3)$ after derivations with respect to x and ξ.*

c) *If V has derivatives of all orders, then for each $M \in \mathbb{N}$*

$$u(t, x, \xi, h) = (2\pi h)^{-n/2} J(t, x, \xi)^{-1/2} \exp\{-S(t, x, \xi)/h\} \times (1 + h\psi_1 + h^2\psi_2 + \cdots + h^M \psi_M + O(h^{M+1} t^{M+3})), \tag{2.6}$$

where the ψ_j are defined by the recursive formulas

$$\psi_j(t, x, \xi) = \frac{1}{2}\int_0^t J^{1/2} \Delta(\psi_{j-1} J^{-1/2})\, d\tau \tag{2.7}$$

(here the integral is taken along the trajectory $X(\tau, \xi, p_0(t, x, \xi))$, while the Laplace operator is applied to the argument x, $\psi_0 = 1$) and satisfy the estimates: $\psi_j = O(t^{j+2})$, $j \geqslant 1$.

To begin with, we shall prove some useful preliminary propositions. In the space of continuous matrix-valued functions on the segment $[0, t]$ we define linear operators G_1, G_2 by the formulas

$$G_1F(t) = \int_0^t d\tau \int_0^\tau V''(x(s))F(s)\,ds,$$
$$G_2F(t) = \int_0^t d\tau\, V''(x(\tau)) \int_0^\tau F(s)\,ds.$$

LEMMA 1. *The norms of the operators G_1, G_2 are bounded by the number $t^2C/2$ and for small $t \leqslant t_0$ are less than one. For such t the derivatives of the solutions of the system* (2.3) *with respect to the initial data are given by the formulas*

$$\frac{\partial X}{\partial x}(t, x, p) = (\mathrm{Id} - G_1)^{-1}E \equiv (\mathrm{Id} + G_1 + G_1^2 + \cdots)E,$$
$$\frac{\partial X}{\partial p}(t, x, p) = (\mathrm{Id} - G_1)^{-1}tE,$$
$$\frac{\partial P}{\partial p}(t, x, p) = (\mathrm{Id} - G_2)^{-1}E,$$
$$\frac{\partial P}{\partial x}(t, x, p) = (\mathrm{Id} - G_2)^{-1}\int_0^t V''(x(\tau))\,d\tau.$$

Here E is the unit $n \times n$ matrix, Id *is the identity operator. In particular,*

$$\frac{\partial X}{\partial p}(t, x, p) = t(E + O(t^2)), \qquad \frac{\partial P}{\partial p}(t, x, p) = E + O(t^2), \tag{2.8}$$

where $O(t^2)$ is uniform with respect to all $t \leqslant t_0$, $x, p \in \mathbb{R}^n$.

The proof is obtained from the standard representation of the solution of the system in variations

$$\dot{A} = \begin{pmatrix} 0 & E \\ V''_{xx}(X(\tau, x, p)) & 0 \end{pmatrix} A, \qquad A(0) = \begin{pmatrix} E & 0 \\ 0 & E \end{pmatrix},$$

which is satisfied by the $2n \times 2n$ matrix $A(t) = \partial(X, P)/\partial(x, p)$ (in the T-exponent series).

We recall that the functions S, P, X defined above satisfy the following well-known identities [7]:

$$\frac{\partial^2 S}{\partial x^2}(t, x, \xi) = \frac{\partial P}{\partial p}(t, \xi, p_0(t, x, \xi))\left(\frac{\partial X}{\partial p}(t, \xi, p_0(t, x, \xi))\right)^{-1}, \tag{2.9}$$
$$\frac{\partial^2 S}{\partial \xi^2}(t, x, \xi) = \left(\frac{\partial X}{\partial p}(t, \xi, p_0(t, x, \xi))\right)^{-1}\frac{\partial X}{\partial x}(t, \xi, p_0(t, x, \xi)). \tag{2.10}$$

Let $t_1 + t_2 \leqslant t_0$. Let us introduce the function

$$f(\eta) = S(t_1, x, \eta) + S(t_2, \eta, \xi).$$

We denote

$$\widetilde{\eta} = X(t_2, \xi, p_0(t_1 + t_2, x, \xi)),$$
$$\widetilde{p}_\eta = P(t_2, \xi, p_0(t_1 + t_2, x, \xi)),$$
$$p_0 = p_0(t_1 + t_2, x, \xi) = p_0(t_2, \widetilde{\eta}, \xi).$$

LEMMA 2. *The following matrix equation is satisfied*:

$$\frac{\partial X}{\partial p}(t_1 + t_2, \xi, p_0) = \frac{\partial X}{\partial p}(t_1, \widetilde{\eta}, \widetilde{p}_\eta) f'' \frac{\partial X}{\partial p}(t_2, \widetilde{\eta}, \xi). \tag{2.11}$$

In particular,

$$\det f'' = J(t_1 + t_2, x, \xi)\, J^{-1}(t_1, x, \widetilde{\eta}) J^{-1}(t_2, \widetilde{\eta}, \xi). \tag{2.12}$$

PROOF. Let us represent the map $X(t_1 + t_2, \xi, p)$ as the composition of two maps:

$$(\xi, p) \mapsto (\eta = X(t_2, \xi, p), p_\eta = P(t_2, \xi, p))$$

and $(\eta, p_\eta) \mapsto X(t_1, \eta, p_\eta)$. Then

$$\frac{\partial X}{\partial p}(t_1 + t_2, \xi, p) = \frac{\partial X}{\partial p}(t_1, \eta, p_\eta) \frac{\partial X}{\partial p}(t_2, \xi, p) + \frac{\partial X}{\partial p}(t_1, \eta, p_\eta) \frac{\partial P}{\partial p}(t_2, \xi, p).$$

For $p = p_0$ we have $\eta = \widetilde{\eta}$, $p_\eta = \widetilde{p}_\eta$. Now we substitute $\partial X/\partial x$, $\partial P/\partial p$ expressed from (2.9), (2.10) in terms of $(\partial^2 S/\partial x^2)(t_1, x, \eta)$ and respectively $(\partial^2 S/\partial \xi^2)(t_2, \eta, \xi)$, into the last formula. This gives (2.11). □

LEMMA 3. *The following estimates are satisfied*

$$\frac{\partial^2 S}{\partial x^2}(t, x, \xi) = \frac{1}{t}(E + O(t^2)), \tag{2.13}$$

$$J(t, x, \xi) = t^n(1 + O(t^2)), \tag{2.14}$$

where $O(t^2)$ *is uniform with respect to all* $t \leqslant t_0$, $x, \xi \in \mathbb{R}^n$;

$$\frac{\partial^m J}{\partial x^{m_1} \partial \xi^{m_2}} = O(t^{n+2}), \qquad m = m_1 + m_2, \tag{2.15}$$

$$\frac{\partial^{m+2}}{\partial x^{m+2}} S(t, x, \xi) = O(t) \tag{2.16}$$

for all $m \geqslant 1$ *for which there exist continuous derivatives of* V *of order* $m+2$. *Moreover* $O(t^{n+2})$ *and* $O(t)$ *are uniform for all* x, ξ *in any compact set* (*and uniform with respect to all* $x, \xi \in \mathbb{R}^n$ *if the derivatives of* V, *mentioned above, are bounded in* $\mathbb{R}^n$).

The proof follows immediately from Lemma 1 and identity (2.9).

Let us note that (2.13) implies

$$S(t, x, \xi) = \frac{(x-\xi)^2}{2t}(1 + O(t^2)) + S(t, \xi, \xi) + \left(\frac{\partial S}{\partial x}(t, \xi, \xi), x - \xi\right).$$

This yields

$$S(t, x, \xi) = \frac{(x-\xi)^2}{2t}(1 + O(t^2)) + O(t)(1 + \|x - \xi\|), \tag{2.17}$$

where $O(t)$ is uniform with respect to ξ from any compact set and for all $x \in \mathbb{R}^n$.

Proof of Theorem 1. We consider the function

$$G(t, x, \xi, h) = (2\pi h)^{-n/2}\varphi(t, x, \xi)\exp\{-S(t, x, \xi)/h\}, \tag{2.18}$$

where $\varphi = J^{-1/2}$. From estimates (2.14), (2.17), we immediately obtain $\lim_{t\to\infty} G(t, x, \xi, h) = \delta(x - \xi)$, i.e., G satisfies the initial condition (2.2). It is well known [7] that $S(t, x, \xi)$ satisfies the Hamilton-Jacobi equation

$$\frac{\partial S}{\partial t} + \frac{1}{2}\left(\frac{\partial S}{\partial x}\right)^2 - V(x) = 0$$

and the function $\varphi = J^{-1/2}$ satisfies the transport equation

$$\frac{\partial \varphi}{\partial t} + \frac{\partial \varphi}{\partial x}\frac{\partial S}{\partial x} + \frac{1}{2}\varphi \Delta S = 0.$$

Substituting G in (2.1), we obtain

$$\left[h\frac{\partial}{\partial t} - \left(\frac{h^2}{2}\Delta - V(x)\right)\right]G = -h^2 F,$$

where

$$F = (1/2)(2\pi h)^{-n/2}\Delta\varphi \exp\{-S(t, x, \xi)/h\}. \tag{2.19}$$

It follows from estimates (2.15) that $\Delta\varphi = O(t^2)\varphi$. Therefore

$$F(t, x, \xi, h) = O(t^2)G. \tag{2.20}$$

Thus G is a formal asymptotic solution of the problem (2.1), (2.2).

We shall show that the exact solution of the problem (2.1), (2.2) is equal to the sum of the convergent series

$$u(t, x, \xi, h) = G(t, x, \xi, h) + \sum_{k=0}^{\infty} h^{k+1}\int_0^t\int_{\mathbb{R}^n} G(t-\tau, x, \eta, h)\mathbb{F}^k F(\tau, \eta, \xi, h)\,d\eta\,d\tau, \tag{2.21}$$

where $\mathbb{F}^k$ denotes the kth degree of the integral operator $\mathbb{F}$ defined by the formula

$$\mathbb{F}\psi(t, x, \xi, h) = \int_0^t\int_{\mathbb{R}^n} F(t-\tau, x, \eta, h)\psi(\tau, \eta, \xi, h)\,d\eta\,d\tau.$$

One can easily see that the series (2.21) is well defined and determines a function of the form $G(t, x, \xi, h)(1 + O(ht^3))$, where $O(ht^3)$ is uniform with respect to all x, ξ from any compact set and all $t \leqslant t_0$. For this purpose one needs to estimate successively the image of the operator $\mathbb{F}$ using (1.20), (2.12), estimates from Lemma 3 and the estimate

$$\int_{\mathbb{R}^n} \exp\{-S(t-\tau, x, \eta) - S(\tau, \eta, \xi)\}\, d\eta = O(1)\left(\frac{\pi\tau(t-\tau)}{t}\right)^{n/2},$$

which follows from the fact that the expression in curly brackets is less than $C(\tau^{-1} + (t-\tau)^{-1})(\eta - \eta_0)^2$ for some $C > 0$, $\eta_0 \in \mathbb{R}^n$. In the same way one verifies the asymptotic properties of the derivatives of the function u which were mentioned in the theorem.

In order to prove that formula (2.21) gives a solution of equation (2.1), note that

$$\left[h\frac{\partial}{\partial t} - \left(\frac{h^2}{2}\Delta - V(x)\right)\right]\mathbb{F}G = -h^2\mathbb{F}F + hF.$$

Thus

$$\left[h\frac{\partial}{\partial t} - \left(\frac{h^2}{2}\Delta - V(x)\right)\right]u = -h^2 F - \sum_{k=1}^{\infty} h^{k+2}\mathbb{F}^k F + \sum_{k=0}^{\infty} h^{k+2}\mathbb{F}^k F = 0.$$

Therefore the representation (2.21) is proved.

To get (2.6), we find first the formal asymptotic WKB solution of problem (2.1), (2.2) of required order. This is done in the usual way (as for instance in [7] for the Schrödinger equation). This gives (2.7). The end of the proof is quite the same as for $M = 1$.

§3. The tunnel canonical operator

The global exponential asymptotics of the Green function of equation (2.1) is determined by means of the tunnel canonical operator introduced here. We shall use some facts concerning symplectic geometry and Lagrange manifolds that can be found, for instance, in the book [7].

Let us recall that a simply connected manifold

$$\Lambda = \{r(\alpha) = (p(r(\alpha)), x(r(\alpha)))\}$$

in $2n$-dimensional phase space $\mathbb{R}^{2n} = \mathbb{R}^n_x \times \mathbb{R}^n_p$ (where $\alpha \mapsto r(\alpha)$ is a smooth map $\mathbb{R}^n \mapsto \mathbb{R}^{2n}$) is said to be a *Lagrange manifold* if the Lagrange brackets vanish on it or, equivalently, if the function

$$S(r) = \int_{r_0}^{r} p\, dx, \qquad r \in \Lambda,$$

is well defined, where the integral is taken along any curve in Λ (r_0 is a fixed point of Λ). This function is called the *generating function* of the Lagrange manifold. If the function $S(r)$ is nonnegative, we call it the *entropy* of the manifold Λ. Let $d\mu = \mu(\alpha)\, d\alpha$ be a measure on Λ. The function

$J = \det(\partial x/\partial \alpha)\mu(\alpha)$ is called the *Jacobian* on Λ. A point $r \in \Lambda$ is called *singular* (or *focal*) if $J(r) = 0$, and *nonsingular* otherwise. We shall suppose in the future that the function S may have only isolated zeros on the set of all singular points. We call r a *nonessential* point of Λ, if there is another point $r' \in \Lambda$ with the same projection on $\mathbb{R}^n_x$ and such that the entropy in it is less than in r.

Now we shall construct an operator K which takes functions on Λ to functions on $\mathbb{R}^n$. This operator will determine the global asymptotics of the fundamental solution of equation (2.1) and some of its other solutions. To begin with, we shall now define the local tunnel operator in a neighborhood of a nonsingular point of Λ.

Let Ω be a nonsingular chart on Λ, i.e., the Jacobian J does not vanish on Ω. Let $D \subset \mathbb{R}^n_x$ denote the set $\pi_x(\Omega^c)$, where π_x is the natural projection on $\mathbb{R}^n_x$ and Ω^c is the closure of the set of all essential points of Ω. Let $D^\gamma \subset \mathbb{R}^n_x$ be the γ-neighborhood of D, i.e., $|x - x'| \geqslant \gamma$ for all $x \in \mathbb{R}^n_x \setminus D^\gamma$, $x' \in D$. Let $\theta(x, \gamma)$ be a smooth function which is equal to one at $x \in D^\gamma$ and vanishes at $x \in \mathbb{R}^n_x \setminus D^{2\gamma}$. Let $\varphi(r, h) \in C_0^\infty(\Omega)$ for all $h \geqslant 0$ (i.e., φ is a finite smooth function). We define the operator $K(\Omega)$ by the formula

$$K(\Omega)\varphi(x) = |J^\gamma|^{-1/2} \exp\{-S^\gamma/h\}\varphi(r(x), h)\theta(x, \gamma), \tag{3.1}$$

where $r(x) \in \Lambda$ is the solution of the equation $x = x(r)$ and J^γ, S^γ are smooth functions which coincide on D with J and S respectively. The function $K(\Omega)\varphi$ certainly depends on γ, $\theta(x, \gamma)$, and in this way the extension of J, S to the functions J^γ, S^γ is carried out.

For singular points formula (3.1) involves singularities, and one cannot use it when calculating asymptotic solutions. In a neighborhood of a singular point, we shall use another formula which is based on different coordinates in phase space. Namely, let I be a subset of $\{1, \dots, n\}$, $H_I = (1/2)\sum_{j\in I} p_j^2$ and let $g_{H_I}^{-\sigma}$ denote the shift of phase space in time σ along the trajectories of the Hamiltonian flow defined by the Hamiltonian function $-H_I$. Let $\sigma > 0$ is small enough that the domain $g_{H_I}^{-\sigma}\Omega$ still has a one-to-one projection onto $\mathbb{R}^n_x$. Then one may take the value of the projection $x(r^\sigma)$ of a point $r^\sigma \in g_{H_I}^{-\sigma}\Omega$ onto $\mathbb{R}^n_x$ as its coordinate:

$$x_I(r^\sigma) = x_I(r) - \sigma p_I, \qquad x_{\bar I}(r^\sigma) = x_{\bar I}(r).$$

Here $\bar I = \{1, \dots, n\} \setminus I$ and $x_M = (x_{j_1}, \dots, x_{j_{|M|}})$ for any subset $M \subset \{1, \dots, n\}$. When shifting a Lagrange manifold along the trajectories of a Hamiltonian flow, one should naturally (from the point of view of Hamiltonian mechanics) change its entropy and Jacobian:

$$S(r^\sigma) = S(r) + \int_0^\sigma (p\,dx + H_I\,dt),$$

where the integral is taken along the trajectory joining the initial point $r \in \Omega$ and the shifted point $r^\sigma(r)$,

$$J(r^\sigma(r)) = \mu(\alpha) \det \frac{\partial x(r^\sigma)}{\partial \alpha}.$$

In our situation

$$S(r^\sigma(r)) = S(r) - \frac{\sigma}{2} p_I^2(r), \qquad J(r^\sigma(r)) = J(r) \det \left(E - \sigma \frac{\partial^2 S_I}{\partial x_I^2} \right).$$

In the new coordinates $x_I(r^\sigma)$, $x_{\bar{I}}(r^\sigma)$ we can try to define the local operator again by formula (3.1), but this formula gives us a function different from the function defined in the initial coordinates. To avoid this difference and to preserve invariance we must apply the operator $\exp\{-\sigma \widehat{H}_I\}$ to this function, where $\exp\{-\sigma \widehat{H}_I\}$ is the solving operator of the Cauchy problem for the k-dimensional heat conduction equation. This procedure leads us to the following definition of the rotated local operator $\widetilde{K}(\Omega)$:

$$\begin{aligned}\widetilde{K}(\Omega)\varphi(x) &= \exp\{-\sigma \widehat{H}_I\} K(g_{H_I}^{-\sigma}\Omega)\varphi(x) \\ &= (2\pi h \sigma)^{-k/2} \int_{\mathbb{R}^k} \exp\left\{ -\frac{|\eta - x_I|^2}{2h\sigma} \right\} (K(g_{H_I}^{-\sigma}\Omega))|_{x_I = \eta}\, d\eta, \quad (3.2)\end{aligned}$$

where the function φ on $g_{H_I}^{-\sigma}\Omega$ is defined by its value on Ω: $\varphi(r^\sigma(r)) = \varphi(r)$, and the essential points on $g_{H_I}^{-\sigma}\Omega$ are regarded as images of the essential points from Ω.

LEMMA. *Let φ and $\widetilde{\varphi}$ be finite smooth functions on Ω such that at essential points of the chart Ω the difference $\varphi(r, h) - \widetilde{\varphi}(r, h)$ is of order $O(h)$. Then at $x \in \Omega^c$*

$$\widetilde{K}(\Omega)\widetilde{\varphi}(x) = K(\Omega)(\varphi + O(h))(x).$$

PROOF. By the definition, we have

$$\begin{aligned}\widetilde{K}(\Omega)\widetilde{\varphi}(x) = (2\pi h\sigma)^{-k/2} \int_{\mathbb{R}^k} & \exp\left\{ -\frac{|\eta - x_I|^2}{2h\sigma} \right\} \\ \times \Bigg[& \exp\left\{ -\frac{1}{h}\left(S - \frac{\sigma}{2}p_I^2\right)(r(r^\sigma)) \right\} \theta(x, \gamma) \\ & \times \varphi(r^\sigma(x)) J(r(r^\sigma(x)))^{-1/2} \det \frac{\partial x(r^\sigma)}{\partial x(r)} \Bigg] \Bigg|_{x_I(r^\sigma) = \eta} d\eta. \quad (3.3)\end{aligned}$$

Here $\theta(x, h)$ is a smooth function with compact support which is equal to one in the γ-neighborhood of $\pi_x(g_{H_I}^{-\sigma}\Omega^c)$. In order to calculate this integral by the Laplace method, let us find the zero points η of the derivative of the preexponential

$$\widetilde{S} = \frac{(\eta - x_I)^2}{2\sigma} + \left(S - \frac{\sigma}{2}p_I^2\right)(r(r^\sigma))|_{x_I(r^\sigma) = \eta}. \qquad (3.4)$$

We have

$$\frac{\partial \widetilde{S}}{\partial \eta} = \frac{\eta - x_I}{\sigma} + \left[\frac{\partial S}{\partial x_I}\left(E - \frac{\sigma}{2}\frac{\partial p_I}{\partial x_I}\right)\left(\frac{\partial x_I(r^\sigma)}{\partial x_I(r)}\right)^{-1}\right]\Bigg|_{x_I(r^\sigma)=\eta}$$
$$= \frac{\eta - x_I}{\sigma} + \frac{\partial S}{\partial x_I}\bigg|_{x_I(r^\sigma)=\eta}.$$

This value vanishes at points η such that

$$\eta = x_I + \sigma \frac{\partial S}{\partial x_I}(r(r^\sigma))|_{x_I(r^\sigma)=\eta}.$$

Consequently, for each $x \in \pi_x \Omega$ there exists a unique η satisfying this equation, namely $\eta(x) = x_I(r^\sigma(r(x)))$. Now let us calculate the second derivative of preexponential (3.4):

$$\frac{\partial^2 \widetilde{S}}{\partial \eta^2} = \frac{1}{\sigma} + \frac{\partial^2 S}{\partial x_I^2}(r(r^\sigma))\left(\frac{\partial x(r^\sigma)}{\partial x(r)}\right)^{-1}\Bigg|_{x_I(r^\sigma)=\eta}$$
$$= \frac{1}{\sigma}\left(\frac{\partial x(r^\sigma)}{\partial x(r)} + \frac{\partial^2 S}{\partial x_I^2}(r(r^\sigma))\sigma\right)\left(\frac{\partial x(r^\sigma)}{\partial x(r)}\right)^{-1}\Bigg|_{x_I(r^\sigma)=\eta}$$
$$= \frac{1}{\sigma}\left(\frac{\partial x(r^\sigma)}{\partial x(r)}\right)^{-1}\Bigg|_{x_I(r^\sigma)=\eta},$$

because

$$\frac{\partial x(r^\sigma)}{\partial x(r)} = E - \sigma \frac{\partial^2 S}{\partial x_I^2}.$$

Thus $\eta(x)$ is proved to be the point of nondegenerate minimum of the function $\widetilde{S}(\eta)$. Substituting the critical value $\eta(x)$ in the formula (3.4) for $\widetilde{S}$, we immediately find: $\widetilde{S}(\eta(x)) = S(x)$.

Now everything has been prepared to write the asymptotics of the integral (3.3) using the Laplace method (for $x \in \pi_x(\Omega^c)$):

$$\widetilde{K}(\Omega)\varphi(x) = (2\pi h\sigma)^{-k/2}(2\pi h)^{k/2}\exp\{-S/h\}\left(\sigma^k \det\left(\frac{\partial x(r^\sigma)}{\partial x(r)}\right)\right)^{1/2}$$
$$\times\, \theta(x(r^\sigma(r(x))), \gamma) J(r(x))^{-1/2}$$
$$\times \det\left(\frac{\partial x(r)}{\partial x(r^\sigma)}\right)^{1/2}(\widetilde{\varphi}(r(x)) + O(h)).$$

After evident simplifications, we get

$$\widetilde{K}(\Omega)\widetilde{\varphi}(x) = \theta(x(r^\sigma(r(x))), \gamma)(\widetilde{\varphi}(r(x)) + O(h))J(r(x))^{-1/2}\exp\{-S/h\}.$$

Therefore, $\widetilde{K}(\Omega)\widetilde{\varphi} = K(\Omega)(\varphi + O(h))$, at all points $x \in \Omega^c$.

This lemma suggests how to define the local operator in a neighborhood of a singular point. As is well known from symplectic geometry [3, 7], in a neighborhood of each such point r on Λ there exists a focal coordinate system $(p_I, x_{\bar{I}})$, where I is a subset of $\{1, \dots, n\}$, $\bar{I} = \{1, \dots, n\} \setminus I$. Moreover, the minimal possible cardinality $|I|$ of the set I is equal to the corank of the matrix $(\partial x(r)/\partial\alpha)$. It is clear that each singular point has a neighborhood Ω in Λ with focal coordinates and a positive number σ such that the transformation $\Omega \mapsto g_{H_I}^{-\sigma}\Omega$ puts Ω in a position which has a one-to-one projection on $\mathbb{R}_x^n$ and the function $\widetilde{S}(\eta)$ of the form (3.4) is nonnegative for all $x \in \pi_x\Omega$. For such a neighborhood Ω, we define the local tunnel operator by formula (3.2). Using arguments similar to those used in the lemma, one may prove the invariance of the local operator thus defined (in the same sense as in the lemma) with respect to the value of the rotation σ, its direction determined by the set I and also with respect to other parameters included in the definition. The global tunnel operator is now defined by means of a suitable partition of unity as in the case of the ordinary canonical operator. Namely: let us choose an atlas of charts on Λ: $\Lambda = \bigcup_{j=1}^{\infty}\Omega_j$ such that for each singular chart Ω_j there exist focal coordinates $(p_I, x_{\bar{I}})$ and a number σ satisfying the conditions pointed out above which are necessary for the definition of the local operator $K(\Omega_j)$. Let $\{e_j(r)\}$ be a partition of unity that corresponds to the chosen atlas. Then the *tunnel canonical operator* is defined as the map $K\colon C_0^\infty(\Lambda \times (0, 1)) \to C^\infty(\mathbb{R}_x^n \times (0, 1))$ by the formula

$$K\varphi = \sum_j K(\Omega_j)(e_j\varphi).$$

Let us introduce the following equivalence relation on $C^\infty(\mathbb{R}_x^n)$: two functions $g, h \in C^\infty(\mathbb{R}_x^n)$ are said to be equivalent if the difference $g - h$ belongs to the space of functions of the form $K\varphi$, where φ is of order $O(h)$ in $\pi_x(\Lambda^c)$. Note that the equivalence of two nonvanishing at $\pi_x(\Lambda^c)$ functions of the form $K\varphi$ and $K\psi$ means that we have the relation $K\varphi/K\psi = 1 + O(h)$. Let Φ denote the corresponding quotient space. The following statement is a consequence of the definition and the lemma.

THEOREM 2. *The tunnel canonical operator*

$$K\colon C_0^\infty(\Lambda \times (0, 1)) \to \Phi$$

does not depend on the chosen canonical atlas, the partition of unity and the parameters included in the definition of the local operator.

REMARK 1. Let Λ^0 be the Lagrange manifold

$$\Lambda^0 = \{(x, p) \in \mathbb{R}^{2n} : x = x(\alpha),\ p = p(\alpha),\ \alpha \in \mathbb{R}^n\}$$

with a positive measure $d\mu(\alpha) = \mu(\alpha)\, d\alpha_1 \cdots d\alpha_n$ and nonnegative entropy S_0, and let $H(x, p)$ be a Hamiltonian function such that the corresponding Lagrange function is also nonnegative. On the Lagrange manifold $\Lambda^t = g_H^t\Lambda^0$ obtained from the initial manifold Λ by means of a shift along the trajectories of the Hamiltonian flow g_H^t, the transferred measure $d\mu_t$ is defined as well as the entropy S^t:

$$S^t(r) = S^0(r_0) + \int_0^r \left(p(\tau)\frac{\partial H}{\partial p}(\tau) - H(p(\tau), x(\tau))\right)d\tau, \tag{3.5}$$

where $r \in \Lambda^t$ is the image of $r^0 \in \Lambda^0$ after its transport along the solutions $p(\tau)$, $x(\tau)$ of the Hamiltonian system and the integral is taken along this solution. It is a well-known fact [3, 7] that the function S^t defined in (3.5) is precisely the generating function of the Lagrange manifold Λ^t.

REMARK 2. In some important situations the tunnel operator is defined not only for finite functions but for all bounded smooth functions in Λ. One can easily see that this is the case when a Lagrange manifold has the following additional property: for each $x \in \mathbb{R}^n$ the set $\{p \in \mathbb{R}^n : (x, p) \in \Lambda\}$ is compact. This condition is satisfied for the manifolds Λ^t, $t > 0$, which are introduced in the next section and play an essential role in the subsequent theory.

§4. Global asymptotics of the fundamental solution and large deviation problems

The tunnel canonical operator defined in the previous section will permit us to write out the exponential asymptotics for the Green function of equation (2.3) at an arbitrary moment of time.

Let $\Lambda^{t,\xi}$ be the Lagrange manifold obtained by the shift of the plane $\Lambda^{0,\xi} = \{(p, x) : x = \xi\}$ along the trajectories of the Hamiltonian flow (2.3) in time t. The entropy on $\Lambda^{0,\xi}$ is considered to be identically zero and the measure on it is one (in the coordinates p). The transport of these objects to the shifted manifold $\Lambda^{t,\xi}$ is described in Remark 1 at the end of the previous section. Let $K^{t,\xi}$ denote the canonical tunnel operator on $\Lambda^{t,\xi}$.

THEOREM 3. *The solution $u(t, x, \xi, h)$ of the problem* (2.1), (2.2) *is defined by the formula*

$$u(t, x, \xi, h) = (2\pi h)^{-n/2}K^{t,\xi}(1 + O(h)). \tag{4.1}$$

In particular, if there exists a finite number k of trajectories $X(\tau, \xi, p_j)$ joining the points ξ, x in time t and such that the functional (2.4) *takes its minimal value $S(t, x, \xi)$ on them, then*

$$u(t, x, \xi, h) = (2\pi h)^{-n/2}\exp\{-S(t, x, \xi)/h\} \times \sum_{j=1}^{k} J_j^{-1/2}(t, x, \xi)(1 + O(h)), \tag{4.2}$$

where J_j is the Jacobian along the trajectory $X(\tau, \xi, p_j)$.

PROOF. If t is small enough, the theorem is the immediate consequence of (2.5) and the definition of the tunnel operator. If $t \in ((m-1)t_0, mt_0]$, the solution of problem (2.1), (2.2) can be written down in the form

$$u(t, x, \xi, h) = \mathbb{G}_\tau^m \delta(x - \xi) = \mathbb{G}_\tau^{m-1} u(\tau, x, \xi, h), \tag{4.3}$$

where $\tau = t/m \in (0, t_0]$ and $\mathbb{G}_\tau^m$ denotes the mth power of the operator

$$\mathbb{G}_\tau \varphi(x, \xi, h) = \int_{\mathbb{R}^n} u(\tau, x, \eta, h)\varphi(\eta, \xi, h)\, d\eta. \tag{4.4}$$

If x is not a focal point of the manifold $\Lambda^{t,\xi}$, then there exists a finite number k of solutions $X_j(\tau) = X(\tau, \xi, p_j)$ of the Hamiltonian system (2.3) such that $X_j(0) = \xi$, $X_j(t) = x$ and the functional (2.4) takes its minimal value on them. It follows from the Laplace method that, up to exponentially small values, the solution (4.3) is equal to the sum of k terms of the form

$$\int_{V_j^1} \cdots \int_{V_j^{m-1}} u(\tau, x, \eta_1, h) \cdots u(\tau, \eta m - 1, \xi, h)\, d\eta_1 \cdots d\eta_{m-1}, \tag{4.5}$$

where V_j^l is a small neighborhood of the point $X_j(l\tau)$ such that for any $y \in V_j^l$ there exists only one trajectory joining ξ and y in time $l\tau$ and on which the functional (2.4) takes its minimal value. We replace all functions u in (4.3) by the asymptotic Green function G. Then the expression (4.3) changes by an exponentially small value. Now we get (4.2) successively calculating all the integrals in (4.5) using the Laplace method and identity (2.12).

If x is the projection of an essential focal point, then the direct application of the Laplace method for the calculation of (4.3) is impossible, because the stationary points of the phase are degenerate. This obstacle will be by-passed in the following way. Using the lemma of §3, let us write the integral kernel of the last operator in (4.3) in the form (3.2), (3.3), where the subset I is chosen in such a way that the revolted neighborhood $g_{H_I}^{-\sigma}\Omega$ of the focal point $r(x)$ in $\Lambda^{t,\xi}$ already has a one-to-one projection on $\mathbb{R}^n$. For example, if $m = 2$, then after this procedure formula (4.3) takes the form

$$\begin{aligned} u(t, x, \xi, h) = (2\pi h\sigma)^{-|I|/2} &\int_{\mathbb{R}^{|I|}} \exp\left\{-\frac{|\eta - x_I|^2}{2h\sigma}\right\} \\ &\times \int_{\mathbb{R}^n} \exp\left\{-\frac{1}{h}(\widetilde{S}(r(r^\sigma)) + S(\tau, z, \xi))\right\}\theta(x, \gamma) \\ &\times J^{-1/2}(r(r^\sigma(x)))J(\tau, z, \xi)^{-1/2}|_{x_I(r^\sigma)=\eta}\, dz\, d\eta, \end{aligned}$$

where $\widetilde{S}(r(r^\sigma))$ is equal to

$$\left(S - \frac{\sigma}{2}\left(\frac{\partial S}{\partial x_I}\right)^2\right)(\tau, x(r(r^\sigma(\eta, x_{\bar I}))), z).$$

Now we get (4.1) by calculating the integral in z by the Laplace method. The theorem is proved.

Now let us find the asymptotic solution of the large deviation problem, i.e., the Cauchy problem for equation (2.1) with initial data

$$u|_{t=0} = \varphi^0(x), \tag{4.6}$$

where $\varphi^0(x)$ is a discontinuous function that does not vanish in some closed domain $D_0 \subset \mathbb{R}^n_x$ with a smooth boundary ∂D_0, $\varphi^0 \in C^\infty(D_0)$ and $\varphi^0(x) \equiv 0$ outside D_0.

Using formula (4.1) for the Green function, we see that the solution $u(t, x, h)$ of the problem (2.1), (4.6) is equal to

$$u(t, x, h) = (2\pi h)^{-n/2} \int_{\mathbb{R}^n} K^{t,\xi}(1 + O(h))\varphi^0(\xi)\, d\xi. \tag{4.7}$$

Evidently, for each t the whole space $\mathbb{R}^n_x$ is naturally divided into two parts In^t and Out^t (perhaps not disjoint) where the minimal point $\xi(t, x)$ of the function $S(t, x, \xi)$ among all $\xi \in D$ belongs respectively to the interior $\mathrm{Int}\, D$ of the domain D and to its boundary. It turns out that the integral (4.7) can be calculated by the Laplace method for all x outside the boundary of the domain In^t. In fact, let $n(x)$ denote the unit outer normal vector to the boundary of the domain D at a point $x \in \partial D$. Let us introduce two Lagrange manifolds in $\mathbb{R}^n_x \times \mathbb{R}^n_p$:

$$\Lambda^0_{\mathrm{in}} = \{x \in \mathbb{R}^n_x,\ p = 0\}, \qquad \Lambda^0_{\mathrm{out}} = \{x \in \partial D,\ p = \rho n(x),\ \rho \in \mathbb{R}\}.$$

The measure on them is considered to be equal to 1 (in Euclidean coordinates x and in coordinates $\rho, \alpha_1, \dots, \alpha_{n-1}$ respectively, where $\alpha_1, \dots, \alpha_{n-1}$ are orthonormal curvilinear coordinates in ∂D) and their entropy vanishes. We denote by K^t_{in} and K^t_{out} the tunnel operators on the Lagrange manifolds $\Lambda^t_{\mathrm{in}} = g^t_H \Lambda^0_{\mathrm{in}}$ and $\Lambda^t_{\mathrm{out}} = g^t_H \Lambda^0_{\mathrm{out}}$ respectively (the transformation of a measure and an entropy by the Hamiltonian flow is described in Remark 1 at the end of §3). Evidently, the set In^t belongs to the projection on $\mathbb{R}^n_x$ of the image of the set $\{x \in D, p = 0\}$ in Λ^t_{in}.

THEOREM 4. *The solution of problem* (2.1), (4.6) *has the form*

$$u(t, x, h) = K^t_{\mathrm{in}}(\varphi^0(g^{-t}_H(r)) + O(h)) \tag{4.8}$$

if $x \in \mathrm{In}^t$ *and*

$$u(t, x, h) = \left(\frac{h}{2\pi}\right)^{n/2} K^t_{\mathrm{out}}(\rho(r)^{-1}\varphi^0(g^{-t}_H(r)) + O(h)) \tag{4.9}$$

if x *belongs to the interior of* Out^t, *where* $\rho(r) = \|p(g^{-t}_H(r))\|$.

The proof is obtained by the direct calculation of the integrals

$$\int_{\mathbb{R}^n} K^{t,\xi}(\Omega_j)(e_j\varphi^0)\, d\xi$$

by the Laplace method, where $\{\Omega_j\}$ is a canonical atlas on $\Lambda^{t,\xi}$ and $\{e_j\}$ is the corresponding partition of unity. We only note that if calculations are to be carried out in a neighborhood of a focal point, one needs to use the method of proof of the previous theorem.

§5. Asymptotic behavior of the solution for the large values of time (of order $h^{-(1+\varkappa)}$, $\varkappa > 0$)

Let the potential V have the additional properties: $V(x) \to \infty$ as $|x| \to \infty$ and $V(x)$ vanishes only at a finite number of points (of global minimum) $\xi_1, \dots, \xi_l$ at which the matrix of second derivatives V'' is not singular. We denote by $(\omega_j^k)^2$ the eigenvalues of the matrix $V''(\xi_k)$, where all ω_j^k are positive. In this section we investigate the asymptotic behavior of the solution of the problem (2.1), (4.6) for large t of order $h^{-(1+\varkappa)}$, $\varkappa > 0$, and also the limits of its logarithmic asymptotics as $t \to \infty$, if the potential V has the properties mentioned above. First of all, we study some properties of the solutions of the Hamilton-Jacobi equation and the Hamilton system in the case of the Hamiltonian $H = p^2/2 - V(x)$.

Let us recall that the resolving operator R_t of the Cauchy problem for the Hamilton-Jacobi equation

$$\frac{\partial S}{\partial t} + \frac{1}{2}\left(\frac{\partial S}{\partial x}\right)^2 - V(x) = 0, \qquad S|_{t=0} = S_0(x) \tag{5.1}$$

is a map $S_0 \mapsto R_t S_0$ such that $S(t, x) = R_t S_0(x)$ is the generalized solution of problem (5.1). The operator R_t is known to be defined on the set of functions bounded from below, by the formula

$$R_t S_0(x) = \inf_{\xi}(S_0(\xi) + S(t, x, \xi)) \tag{5.2}$$

(see, for instance, [9]). The semigroup of the nonlinear operators R_t is uniquely defined by the following properties:

a) the function $R_t S_0$ gives the classical (i.e., everywhere smooth) solution of problem (5.1) if t is small and the initial function S_0 is smooth and convex; b) the operators R_t are continuous in some natural topology on the space of all functions bounded from below; c) the operators R_t commute with the operation of taking the minimum of two functions and with the operation of augmentation of the function by a constant, i.e.,

$$\begin{gathered} R_t(\min(S_1, S_2)) = \min(R_t S_1, R_t S_2), \\ R_t(\lambda + S(x)) = \lambda + R_t S(x), \qquad \lambda \in \mathbb{R}. \end{gathered} \tag{5.3}$$

Moreover, the image R_t of any function satisfies the Hamilton-Jacobi equation at all the points of its differentiability. It follows from (5.2) that this image satisfies local Lipschitz condition and therefore it is differentiable almost everywhere.

Let us recall that the function $S(t, x, \xi)$ from formula (5.2) was introduced at the beginning of §2 as the infimum of the functional

$$J(y(\cdot)) = \int_{-t}^{0} \left(\frac{1}{2}\dot{y}^2(\tau) + V(y(\tau))\right) d\tau, \tag{5.4}$$

defined on the set of all continuous piecewise smooth functions on the segment $[-t, 0]$ with fixed endpoints $y(-t) = \xi$, $y(0) = x$. The functional (5.4) is proved to take its minimal value on a smooth curve [6]. This fact follows because it is true for small t (less than some fixed t_0) and the minimum of (5.4) among piecewise smooth curves for any t coincides with the minimum among those curves whose derivatives are continuous with only possible exceptions at the points kt_0, where $k \in \mathbb{N}$, $k \leqslant t/t_0$. We also note the symmetry of the function S: $S(t, x, y) = S(t, y, x)$ for all t, x, y, which follows from the invariance property of the Hamiltonian system (2.3) with respect to inversion of time.

Now let us define the functions

$$S_k(x) = \inf\{S(t, x, \xi_k) : t > 0\} \tag{5.5}$$

for $k = 1, \dots, l$ and their minimum

$$S(x) = \min\{S_k(x) : k = 1, \dots, l\}. \tag{5.6}$$

It is clear that $S(t, x, \xi_k)$ as a function of t is monotone nonincreasing for each x, ξ_k. Consequently,

$$S_k(x) = \lim_{t\to\infty} S(t, x, \xi_k). \tag{5.7}$$

Moreover, the functions S_k are nonnegative everywhere and $S_k(x) = 0$ only for $x = \xi_k$. First of all, we need some simple facts about these functions.

Proposition 1. *The functions $S_k(x)$, $k = 1, \dots, l$, and $S(x)$ are fixed points of the resolving operator* (5.2) *of the Cauchy problem* (5.1).

Proof. From the definition (5.2) of the operator R_t and the monotonicity of $S(t, x, \xi_k)$, we get successively

$$\begin{aligned} R_t S_k(x) &= \inf_{\xi} \inf_{\tau>0} (S(\tau, \xi, \xi_k) + S(t, x, \xi)) \\ &= \inf_{\tau>0} S(t+\tau, x, \xi_k) = \inf_{\tau>0} S(\tau, x, \xi_k) = S_k(x). \end{aligned}$$

Now (5.3) implies the immobility of $S(x)$ under the action of R_t.

Proposition 2. *Let x satisfies the inequality*

$$S_k(x) < \min_{j\neq k}(S_j(x) + S_k(\xi_j)). \tag{5.8}$$

Then there exists a smooth curve $q\colon (-\infty, 0] \to \mathbb{R}^n$ such that

$$S_k(x) = \int_{-\infty}^{0} \left(\frac{1}{2}\dot{q}^2(t) + V(q(t))\right) dt, \tag{5.9}$$

$$q(0) = x, \quad \lim_{t\to-\infty} q(t) = \xi_k, \quad \lim_{t\to-\infty} \dot{q}(t) = 0.$$

PROOF. Let $k \in \{1, \dots, l\}$, $x \in \mathbb{R}^n$, satisfy (5.8). It follows from the properties of the functions $S_k(x)$, $S(t, x, \xi)$ mentioned above that there exists a sequence $\{q_m(t), t \in [-T_m, 0]\}$ of the solutions of the Hamiltonian system (2.3) with endpoints $q_m(-T_m) = \xi_k$, $q(0) = x$ such that $T_m \to \infty$ as $m \to \infty$ and the functional (5.4) takes its minimal value on q_m, i.e., $J(q_m) = S(T_m, x, \xi_k)$. We shall prove the existence of the limit trajectory in several steps.

A) For each $j \neq k$ there exists a neighborhood U_j of the point ξ_j such that all trajectories q_m except a finite number (by taking a subsequence, we shall suppose further that all of them) have no intersections with U_j. Indeed, assume the contrary. Let us take any $\varepsilon < 0$. Due to the continuity of S_j, we may choose a neighborhood U_j of the point ξ_j, in which $S_j < \varepsilon$. By assumption, there exist trajectories with arbitrary large numbers m such that $\exists t_m : y_m = q_m(-t_m) \in U_j$. Note that $S(t, x_1, x_2) \geqslant S_j(x_1) - S_j(x_2)$ for all $x_1, x_2 \in \mathbb{R}^n$, $j \in \{1, \dots, l\}$. In fact, one can get this inequality by passing to the limit as $s \to \infty$ in the evident inequality

$$S(t, x_1, x_2) \geqslant S(t+s, x_1, \xi_j) - S(s, x_2, \xi_j).$$

Hence

$$\begin{aligned} S(T_m, x, \xi_k) &= S(t_m, x, y_m) + S(T_m - t_m, y_m, \xi_k) \\ &\geqslant S_j(x) + S_j(\xi_k) - 2S_j(y_m) \geqslant S_j(x) + S_j(\xi_k) - 2\varepsilon. \end{aligned}$$

By passing to the limit as $T_m \to \infty$, we get $S_k(x) \geqslant S_j(x) + S_j(\xi_k)$, since ε was arbitrary. But this inequality contradicts (5.8).

B) For each neighborhood U of the point ξ_k such that $x \notin U$ there exist an $N \in \mathbb{N}$ and moments of time $t_1 < t_0 < 0$ such that for all $m > N$ we have

$$t_m(U) = \sup\{\tau : q_m(\tau) \in U\} \in [t_1, t_0].$$

Indeed, the value of the functional (5.4) on the part of a trajectory that remains outside of U during the period of time t is greater than

$$t \min\{V(y) : y \notin U, \forall j \neq k, y \notin U_j\},$$

tends to infinity as $t \to \infty$, and therefore this period of time must be bounded. On the other hand, "the time of movement" from x to U cannot be too small, since $\min_{y \in U} S(\tau, x, y) \to \infty$ as $\tau \to 0$.

C) It follows from statement B) that the sequences $y_m(U) = q_m(t_m(U))$ and $t_m(U)$ belong to the compact sets ∂U and $[t_1, t_0]$ respectively. Therefore these sequences have limit points and one can choose converging subsequences. Let B_p be the sequence of balls of radius $1/p$ and center ξ_k. Using the diagonal method, let us construct now the subsequence of trajectories $q_m(\tau)$ such that the sequences $t_m(B_p)$ and $y_m(B_p)$ have limits $t(p)$ and $y(p)$ respectively. Now the Arzelà theorem permits to choose a subsequence of trajectories such that the sequence of the trajectory segments

$q_m(\tau)$, $\tau \in [t(p), 0]$, converges in C^1-topology for each $p \in \mathbb{N}$. The union of these limit segments gives us the required limit trajectory $q\colon (-\infty, 0] \to \mathbb{R}^n$.

COROLLARY. *There exists a limit*

$$\lim_{t\to\infty} S(t, x, y) = \min_{j,k}(S_k(x) + S_j(y) + S_k(\xi_j)) \tag{5.10}$$

for all points x, y.

Let us discuss the geometric meaning of the functions $S_k(x)$. Since $z_k = (\xi_k, 0)$ is a hyperbolic nondegenerate singular point of the Hamiltonian system, the general theory [11] implies the existence of stable and unstable invariant manifolds W_k^{in} and W_k^{out} embedded in $\mathbb{R}^{2n}$. The dimension of each of them is equal to n, since one half of the eigenvalues of the linear part of the Hamiltonian system in a neighborhood of z_k is positive and the other one is negative. By definition, the *stable* (*unstable*) *manifold* is the set of points in $\mathbb{R}^n$ such that the trajectories starting at these points tend to z_k as $t \to \infty$ (as $t \to -\infty$). In our situation both of these manifolds are Lagrangian, for they are invariant with respect to the Hamiltonian flow, which is a one-parameter group of symplectic transforms (in fact, the value of the symplectic form on a pair of tangent vectors at a point of W_k^{out} or W_k^{in} is invariant and at the same time tends to zero as $t \to \mp\infty$ respectively).

PROPOSITION 3. *In the domain defined by inequality* (5.8), *the function* $S_k(x)$ *is the generating function* (*or the entropy*) *of the essential part of the unstable Lagrange manifold* W_k^{out} *vanishing at* $z_k = (\xi_k, 0)$ (*the definition of the essential part of a Lagrange manifold was given in* §3).

PROOF. The generating function (vanishing in z_k) of the manifold W_k^{out} is defined by the formula

$$\sigma(z) = \int_{z_0}^{z} p\,dq,$$

where the integral is taken along any curve in W_k^{out} joining z_k and $z = (x, p)$. Taking the integral along the trajectory $z(t) = (x(t), p(t))$ of the Hamiltonian system with endpoints $z(-\infty) = z_k$, $z(0) = z$ (such a trajectory exists by the definition of W_k^{out}), we see that

$$\sigma(z) = \int_{-\infty}^{0} \left(\frac{1}{2}\dot{x}^2(t) + V(x(t))\right) dt,$$

which completes the proof.

This implies, in particular, the smoothness of the function S_k in a neighborhood of ξ_k.

PROPOSITION 4. *There exists a measure* $d\mu$ *on* W_k^{out} *which is connected with the Hamiltonian flow by the condition*

$$d\mu(z_t) = \exp\left\{\sum_{j=1}^{n} \omega_j^k t\right\} d\mu(z_0), \tag{5.11}$$

where z_t *is the image of a point* $z_0 \in W_k^{\text{out}}$ *under the action of the Hamiltonian flow* (2.3).

PROOF. The existence of such a measure in a neighborhood of a nondegenerate singular point follows from the theory of normal forms in ordinary differential equations. In the nonresonance case, i.e., if each ω_j^k (let us recall that $(\omega_j^k)^2$ are the eigenvalues of the matrix $V''(\xi_k)$) is not equal to any linear combination of the form

$$\sum_{i\neq j} \nu_i \omega_i^k, \quad \text{where } \sum_{i\neq j} \nu_i \geqslant 2, \ \nu_i \in \mathbb{N},$$

then the system (2.3) can be linearized [11, 15]. This means that in a neighborhood of a singular point in W_k^{out} there exist smooth coordinates $\mu = (\mu_1, \dots, \mu_n)$ in which the system (2.3) becomes linear:

$$\dot{\mu}_j = \omega_j^k \mu_j, \qquad j = 1, \dots, n,$$

and has a solution of the form

$$\mu_j(t) = \exp\{\omega_j^k t\}\mu_j(0), \qquad j = 1, \dots, n.$$

Then, evidently,

$$\det \frac{\partial \mu(t)}{\partial \mu(0)} = \exp\left\{ t \sum_{j=1}^{n} \omega_j^k \right\}, \tag{5.12}$$

and in a neighborhood of z_k the coordinates μ define a measure $d\mu$ on W_k^{out} that satisfies (5.11). It is clear that this measure can be extended to the whole manifold W_k^{out} preserving property (5.11).

If resonance relations exist, the system cannot be linearized by a smooth transformation (in general there exists only continuous local homeomorphism which linearizes system (2.3) on W_k^{out} [11]). But there always exists the Poincaré polynomial normal form, [28] for system (2.3). One can verify that (5.12) holds for the coordinates which define this normal form and consequently, these coordinates define a measure that satisfies (5.11).

It is clear that the measure on W_k^{out} satisfying (5.11) is defined uniquely up to a real multiplier. It will be convenient for us to fix this measure by the following normalization condition: at the singular point $z = (\xi_k, 0)$

$$\det(\partial \mu / \partial x) = 1. \tag{5.13}$$

REMARK. If the potential V is analytic then one can calculate the coordinates μ by means of Dirichlet series [2, 5].

Now we can define the Jacobian

$$J_k = \det(\partial x / \partial \mu) \tag{5.14}$$

at nonsingular points of W_k^{out}. Let us introduce the neighborhood D^k of a point ξ_k, which is the set of all points x such that

$$S_k(x) = \min_j \{S_j(x)\}. \tag{5.15}$$

Evidently, $x \in D^k$ satisfies (5.8). Let Ω^k be a smooth Lagrange manifold (with boundary) such that $\Omega^k \subset W_k^{\text{out}}$, Ω^k contains all essential points of W_k^{out} with projections in a γ-neighborhood of D^k (γ is an arbitrary positive parameter) and does not contain points with projections outside 2γ-neighborhood of D^k. Let us denote by K_{st}^k the tunnel canonical operator on Ω^k with phase S_k and Jacobian J_k defined above. This operator will play a crucial role in the sequel.

Now let us return to the Cauchy problem (2.1), (4.6).

Proposition 5. *Let the potential V have the properties mentioned at the beginning of the section and let the initial function φ^0 in* (4.6) *be the characteristic function of a bounded domain D, with all $\xi_1, \dots, \xi_l$ belonging to the interior of D*:

$$u|_{t=0} = \varphi^0(x) = \begin{cases} 1 & \text{if } x \in D, \\ 0 & \text{if } x \notin D. \end{cases} \tag{5.16}$$

Then there exists a limit, as $t \to \infty$, of the first term of the logarithmic asymptotics for the problem (2.1), (5.16). *This limit does not depend on* D *and has the form*

$$\lim_{t\to\infty} \lim_{h\to 0} (h \log u) = -S(x). \tag{5.17}$$

Proof. It follows from Theorem 4 that $\lim_{h\to 0}(h\log u) = -\sigma(t, x)$, where

$$\sigma(t, x) = \int_0^t \left(\frac{1}{2}\dot{y}^2(\tau) + V(y(\tau))\right) d\tau, \tag{5.18}$$

and $y(\tau) = X(\tau, \xi(t, x), p_0(t, x))$ is a trajectory of the Hamiltonian system (2.3) joining the points $\xi(t, x)$ and x, satisfying $\dot{y}|_{\tau=0} = 0$ (if $\xi(t, x) \in D \setminus \partial D$), $\dot{y}|_{\tau=0}$ being normal to the boundary of D (if $\xi(t, x) \in \partial D$).

By definition, $0 \leqslant \sigma(t, x) \leqslant \min_k S(t, x, \xi_k)$, and consequently

$$\overline{\lim}\, \sigma(t, x) \leqslant S(t). \tag{5.19}$$

Let us denote by $\tilde{\xi}(t, x)$ the point on the trajectory $y(\tau)$ (defined above via (5.18)) where the potential $V(x)$ takes a minimal value and let $\tilde{\xi}(t, x) = y(\tilde{t})$. It is clear that $\sigma(t, x) \geqslant tV(\tilde{\xi}(t, x))$. Consequently, the point $\tilde{\xi}(t, x)$ tends to the point ξ_k of global minimum of V as $t \to \infty$. In particular,

$$\lim_{t\to\infty} S(t, \tilde{\xi}(t, x), \xi_k) = 0.$$

Taking into consideration the inequality

$$\begin{aligned} &S(t, \tilde{\xi}(t, x), \xi_k) + \sigma(t, x) \\ &\quad = S(t, \tilde{\xi}(t, x), \xi_k) + \left(\int_0^{\tilde{t}} + \int_{\tilde{t}}^t\right)\left[\frac{1}{2}\dot{y}^2(\tau) + V(y(\tau))\right] d\tau \\ &\quad \geqslant S_k(2t - \tilde{t}, x, \xi_k) \geqslant S_k(x), \end{aligned}$$

we conclude that

$$\lim_{t\to\infty} \sigma(t, x) \geqslant S_k(x) \geqslant S(x). \tag{5.20}$$

Inequalities (5.19), (5.20) prove the proposition.

Using the corollary to Proposition 2, one can similarly obtain the limit of the logarithmic asymptotic first term for the Green function of equation (2.1).

PROPOSITION 6. *We have*

$$\lim_{t\to\infty}\lim_{h\to 0}(h\log u(t, x, \xi, h)) = -\min_{j,k}(S_k(x) + S_j(\xi) + S_k(\xi_j)),$$

where $u(t, x, \xi, h)$ *is the solution of problem* (2.1), (2.2).

Let us calculate the principal term of the exponential asymptotics of the solution of problem (2.1), (5.16).

PROPOSITION 7. *If the conditions of Proposition* 5 *are satisfied and* x *belongs to the interior of* D^k, *then*

$$\lim_{t\to\infty}\left[\exp\left\{\frac{t}{2}\sum_{j=1}^{n}\omega_j^k\right\}u_0\right] = K_{\mathrm{st}}^k(1 + O(h)), \tag{5.21}$$

where u_0 *is the principal term of the exponential asymptotics for problem* (2.1), (5.16) *and* K_{st}^k *is the tunnel canonical operator introduced above.*

PROOF. By Proposition 5, the solution of problem (2.1), (5.16) for large t and arbitrary x is given by formula (4.8). If $x \in D^k$ is the projection of only nonsingular essential points of the Lagrange manifold Ω^k, then for large t the point x is the projection of only nonsingular points of the Lagrange manifold Λ_{in}^t, that determines the tunnel operator K_{in}^t in formula (4.8). In this case, the Jacobians on Ω^k and Λ_{in}^k above x do not vanish, and the projection of the manifold Ω^k on $\mathbb{R}_x^n$ is a local diffeomorphism near every point of the trajectory joining $z_k = (\xi_k, 0)$ to $z = (x, p)$ in Ω^k. Thus

$$\frac{\partial X}{\partial X(0)}(t) = \frac{\partial x}{\partial \mu}(X(t))\,\frac{\partial \mu}{\partial \mu_0}(t)\,\frac{\partial \mu}{\partial x}(X(0)).$$

Hence, taking into consideration (5.12), (5.13), we get

$$\det\frac{\partial X}{\partial x}(t) = \det\frac{\partial x}{\partial \mu}\exp\left\{t\sum_{j=1}^{n}\omega_j^k\right\}(1 + O(t^{-1})). \tag{5.22}$$

This proves (5.21) for nonsingular points of Ω^k. To prove (5.21) in the general case, one needs to use the method which was applied at the end of the proof of Theorem 3.

COROLLARY. *If the sums* $\sum_{j=1}^{n} \omega_j^k$ *do not depend on* $k = 1, \dots, l$ *and are equal to* $\mathscr{E}$, *then*

$$\lim_{t\to\infty} \exp\{t\mathscr{E}/2\}u_0 = \sum_{k=1}^{l} K_{\mathrm{st}}^k(1 + O(h)). \tag{5.23}$$

We have investigated the behavior of the principal term of the exponential asymptotics for problem (2.1), (5.16). However, to apply this to the Schrödinger operator theory, one needs to know the behavior of the solution itself (not only its principal asymptotics) for large t. The following theorem, which states that the behavior of the solution of problem (2.1), (5.16) and of the principal term of its exponential asymptotics coincide when t is of order $h^{-(1+\varkappa)}$, $\varkappa > 0$, is one of the main results of this paper.

THEOREM 5. *Suppose the sums* $\sum_{j=1}^{n} \omega_j^k$ *do not depend on* k *and are equal to* $\mathscr{E}$. *Then the solution* $u(t, x, h)$ *of the Cauchy problem* (2.1), (5.16) *has the form*

$$u = \exp\{-t\mathscr{E}/2\} \sum_{k=1}^{l} K_{\mathrm{st}}^k(1 + O(h)) \tag{5.24}$$

as $t = h^{-(1+\varkappa)}$, $h \to 0$, *where* $O(h)$ *is uniform with respect to all* x *and* $\varkappa \in (0, 1)$ *from any compact set.*

PROOF. We shall consider only the case when x is the projection of only nonsingular essential points of Ω^k. The general situation can be treated as usual by the method applied at the end of the proof of Theorem 3. It was pointed out already, that the solution $u(t, x, h)$ is defined by formula (4.8) when t is large enough. At the same time,

$$u(t, x, h) = (2\pi h)^{-n/2} \int_{D_0} u(t, x, \xi, h)\, d\xi,$$

where $u(t, x, \xi, h)$ is defined by (4.3).

The main idea of the proof is the following. The minimal possible number of iterations m in (4.3) is t/t_0. But one must take a greater number in order to use the fact that for small t the correction in formula (2.5) has the form $O(ht^3)$ (and not simply $O(h)$). We will take the number of iterations equal to $m = t^{1+\alpha}$, $\alpha > 0$ (more precisely, the integer part of this value). Then the time of one iteration in (4.3) will be of order

$$\tau = t/t^{1+\alpha} = t^{-\alpha} = h^{\alpha(1+\varkappa)}. \tag{5.25}$$

Let $q_r(\tau)$, $\tau \in (-\infty, 0]$, $r \leqslant \tilde{r}$, be a finite number of trajectories joining ξ_k to x in infinite time and with the action $S_k(x)$. We choose a small ε such that in the ε-neighborhood of each trajectory q_r there exists only one minimal trajectory joining ξ_k to x in time t, when t is sufficiently large. Then the fundamental solution is equal to the sum of $\tilde{r}$ terms of the form

(4.5) (with balls of radius ε for V_j) and an exponentially small remainder. Note that one needs to show also that this remainder is uniformly small with respect to t and m.

First let us consider the term (4.5) with $u(t, x, \xi, h)$ of the form (2.5), and V_j being balls of radius ε. We shall calculate the integrals in (4.3) successively, applying the Laplace method with the estimate (1.21) of the remainder. From the estimate of the coefficient $\alpha(h)$ in (1.21) and the estimates of the derivatives of the Jacobian and phase for small τ (see §2) we see that the first three terms in $\alpha(h)$ have the form $O(\tau^3)f(x_0)$. Taking δ_1 small, we reduce the fourth term to the same form. To estimate the last term in (1.21), we must estimate integrals of the form

$$\int_{V_j} \exp\left\{-\frac{1}{h}[S(T, x, \eta_j) + S(\tau, \eta_j, \eta) - S(T+\tau, x, \eta)]\right\} d\eta_j, \tag{5.26}$$

where $T = k\tau$ for some integer k. Note that the function $S(t, x, \eta_j)$ is convex with respect to η_j in V_j. At the same time there is an estimate for the second derivative of the function $S(\tau, \eta_j, \eta)$. Therefore, the expression in braces in (5.16) is not greater than $-(h\tau)^{-1}(1+O(\tau))(\eta_j - \eta_j^0)^2$. Thus the integral (5.26) is bounded from above by the value

$$[2\pi h\tau(1+O(\tau))]^{n/2}. \tag{5.27}$$

Consequently, each application of the Laplace method (each iteration) will add to the principal term of the asymptotics a multiplier of the form $(1+O(h\tau^3))$ and the whole integral (4.5) will differ from this principal term by the factor $\gamma(h)$ which is the product of m such multipliers. Thus, the modulus of the deviation of $\gamma(h)$ from one can be estimated by the value $(1+Ch\tau^3)^m - 1$, where C is a constant, and therefore

$$\begin{aligned}|\gamma(h) - 1| &\leqslant (1 + Ch^{1+3\alpha(1+\varkappa)})^{h^{-(1+\varkappa)(1+\alpha)}} - 1 \\ &= \exp\{h^{-(1+\varkappa)(1+\alpha)} \log(1 + Ch^{1+3\alpha(1+\varkappa)})\} - 1 \\ &= \exp\{O(h^{1+3\alpha(1+\varkappa)-(1+\varkappa)(1+\alpha)})\} - 1 = O(h^{1+(1+\varkappa)(2\alpha-1)}).\end{aligned}$$

Evidently, this expression has the form $O(h)$ if $\alpha > 1/2$.

Now let us deal with the difference between (4.3) and the sum of $\tilde{r}$ terms of the form (4.5). Further on this sum will be called the principal part of (4.3). Let us explain first where the difficulty lies.

When integrating over the exterior of the ε-neighborhood of the trajectories $q_r(\tau)$, the minimum of the phase will exceed $S_k(x)$ by some value Δ, and the multiplier $\exp\{-\Delta/h\}$, which was absent in the principle part, will appear. However, each application of the operator $\mathbb{G}_\tau$ in (4.3) will add the factor $(2\pi h)^{-n/2}$ and after m applications the coefficient $(2\pi h)^{-mn/2}$ appears. This coefficient is not supressed by the multiplier $\exp\{-\Delta/h\}$ if m is of order $h^{-(1+\varkappa)}$, $\varkappa > 0$. In the case of fixed m or m of the order $h^{-(1-\varkappa)}$,

$\varkappa > 0$, the coefficient $(2\pi h)^{-mn/2}$ is small as compared with $\exp\{-\Delta/h\}$ and in that situation it was evident that the difference between (4.3) and its principal part was exponentially small (which we used when proving Theorem 3). In our present situation we need more delicate arguments.

The nonsingularity of the matrix of second derivatives of the potential in a neighborhood of its minimum points implies the existence of neighborhoods U_k of each point ξ_k such that for all pairs $x, y \in U_k$ and $t > 0$ there exists a unique trajectory of the Newton system joining x to y in the time t giving the minimal value of functional (5.4). Moreover, the function $S(t, x, y)$ is smooth and convex in each argument $x, y \in U_k$ for all t. First of all, let us estimate the difference between (4.3) and its principal part for points $x \in U_k$. In this case there exists only one trajectory joining ξ_k to x in infinite time. Integrating over the exterior of its ε-tubular neighborhood we decompose this set in two parts: M_1 is the set of all $\eta_1, \dots, \eta_m$ that belong to U_k and M_2 is the set of all $\eta_1, \dots, \eta_m$ such that at least one of them does not belong to U_k. The first integral can be estimated in the same way as the integral over the interior of the ε-tube of the limit trajectories since, by the convexity of $S(t, x, y)$ for $x, y \in U_k$, integrals of type (5.26) over U_k can be estimated by the expressions (5.27). The second integral we represent as the sum of m integrals over the sets $\Omega_1, \dots, \Omega_m$, where

$$\Omega_j = \{\eta_1, \dots, \eta_{j-1} \in U_k;\ \eta_j \notin U_k;\ \eta_{j+1}, \dots, \eta_m \in \mathbb{R}^n\}.$$

The integral over each Ω_j is exponentially small as compared with the principal term. To prove this, one needs to integrate over $\eta_1, \dots, \eta_{j-1}$ by the previous scheme and then to estimate the obtained phase

$$S(j\tau, \eta_j, \xi) + S(\tau, \eta_{j+1}, \eta_j) + \cdots + S(\tau, x, \eta_m) - S(t, x, \xi)$$

from below by the expression

$$\Delta + \frac{(\eta_{j+1} - \eta_j)^2}{2\tau} + \cdots + \frac{(x - \eta_m)^2}{2\tau},$$

where $\Delta > 0$ is the estimate of the expression $S(j\tau, \eta_j, \xi) - S(t, x, \xi)$ from below, for example $\Delta = \min_{y \notin U_k} S_k(y) - S_k(x)$. In this way we get a sum of $m = h^{(1+\varkappa)(1+\alpha)}$ terms which are exponentially small in h and therefore this sum is also exponentially small.

Now let $x \notin U_k$ and let t be large enough. Then almost all critical points η_j^0 of the phase belong to U_k with the exception of a small part of them of order $h^{-\alpha(1+\varkappa)}$. As it was pointed out above, the estimate is trivial for $h^{-\alpha}$ critical points, if $\alpha < 1$. For other integrals we may use the arguments that was applied in the case $x \in U_k$. Thus, the parameter α must belong to the interval $(1/2, (1+\varkappa)^{-1})$. The proof is complete.

In an entirely analogous way one may prove the following

THEOREM 6. *Let V_k be a closed neighborhood of ξ_k, contained in the interior of the set D^k and not containing other points ξ_j, $j \neq k$; let the assumptions of Theorem 5 hold. If $t = h^{-(1+\varkappa)}$, $\varkappa \in (0, 1)$ and $h \to 0$, then the solution $u_k(t, x, h)$ of the Cauchy problem for equation* (2.1) *whose initial function is the characteristic function of V^k has the form*

$$u_k = e^{-t\mathscr{E}/2} K_k(1 + O(h)) = \exp\{-(\mathscr{E}/2)h^{-(1+\varkappa)}\} K_k(1 + O(h)) \qquad (5.28)$$

up to a term which is exponentially small in $h \to 0$ as compared with the function (5.24).

§6. Exponential asymptotics for low levels of the Schrödinger operator

We consider the stationary Schrödinger equation

$$\widehat{H}\psi = E\psi, \qquad \psi \in L_2(\mathbb{R}^n_x), \ x \in \mathbb{R}^n_x, \qquad (6.1)$$

where

$$\widehat{H} = -(h^2/2)\Delta + V(x), \qquad (6.2)$$

and the potential $V(x)$ satisfies the conditions mentioned at the beginning of §5. We shall also assume that the value $\mathscr{E} = \sum_{j=1}^n \omega_j^k$ does not depend on the point ξ_k, and therefore Theorems 5 and 6 hold. It is well known [14, 24] that the spectrum of the operator (6.2) is discrete if the potential V tends to infinity as $x \to \infty$. Since V is nonnegative, the operator (6.2) is also nonnegative. Due to the localization principle, the power asymptotics of the low levels of $\widehat{H}$ is defined by the behavior of $V(x)$ in a neighborhood of minimum points $\xi_1, \dots, \xi_l$. In particular, $\widehat{H}$ has exactly l eigenvalues of the form $h((\mathscr{E}/2) + O(h))$ and for all other levels E the inequality $E/h - \mathscr{E}/2 \geqslant C$ holds, where C is a positive constant. One can find the proof of these facts, for instance, in [18, 25, 35]. Let ψ_j, $j = 1, \dots, l$, be the eigenfunctions of the operator $\widehat{H}$ related to the levels E_j. To construct the exponential asymptotics of the functions ψ_j, it is convenient to introduce the space Φ, defined as the quotient space of $L_2(\mathbb{R}^n_x)$ by the subspace

$$P = \left\{ f \in L_2(\mathbb{R}^n_x) : f = \sum_{j=1}^l K_j(O(h)) \right\},$$

where the tunnel operators K_j are the same as in Theorems 5 and 6. For brevity we shall also denote by K_j the projections of the functions $K_j(1 + O(h))$ on Φ. The following statement is the basis for the derivation of all formulas obtained further in this section.

THEOREM 7. *The two l-dimensional vector spaces generated by the functions K_j, $j = 1, \dots, l$, and by the projections of the eigenfunctions ψ_j on*

Φ respectively coincide. In particular, given $j = 1, \dots, l$, there exist coefficients $C_i(h)$ such that

$$\psi_j = \sum_{i=1}^{l} C_i(h) K_i$$

up to functions from the space Φ.

PROOF. Let $u(t, x, h)$ be the solution of the heat-conductivity equation (2.1) with initial data $u_k(0, x, h)$ of the form (5.12) and let

$$u_k(0, x, h) = \sum_{j=1}^{\infty} \alpha_j \psi_j(x)$$

be the expansion of the initial function in the sum of eigenfunctions of the operator $\widehat{H}$. Then

$$u_k(t, x, h) = \sum_{j=1}^{\infty} \alpha_j \exp\{-tE_j/h\} \psi_j(x)$$

is evidently the expansion of the solution $u_k(t, x, h)$. Then, putting $t = h^{-(1+\varkappa)}$ and using (5.28), we get

$$K_k(1 + O(h)) = \sum_{j=1}^{\infty} \alpha_j \exp\{-h^{-(1+\varkappa)}(E_j/h - \mathscr{E}/2)\} \psi_j.$$

It follows from the properties of the eigenfunctions mentioned above that, given any compact set in $\mathbb{R}^n_x$, the restrictions of the functions ψ_j with numbers $j > l$ belong to P for small h and therefore the equation

$$K_k = \sum_{j=1}^{l} \beta_j(h) \psi_j$$

holds in Φ for some constants $\beta_j(h)$. This fact and the linear independence of the functions K_j imply the statement of Theorem 7.

It is especially interesting to study problems in which the tunnel effects are essential from the physical point of view. We mean the spectral problems for the Schrödinger operator with potential $V(x)$ being invariant with respect to the action of a nontrivial discrete group of linear transformations of the coordinate space $\mathbb{R}^n$. For these problems it is sometimes possible to calculate explicitly the coefficients $C_j(h)$ from Theorem 7. To demonstrate our methods, we shall consider here two of the simplest examples.

PROPOSITION 1. *Let $V(x)$ vanish only at the points*

$$\xi_1 = (\|\xi_1\|, 0, \dots, 0), \qquad \xi_2 = (-\|\xi_1\|, 0, \dots, 0)$$

and be even with respect to the first coordinate: $V(R(x)) = V(x)$, *where* $R(x_1, x_2, \dots, x_n) = (-x_1, x_2, \dots, x_n)$. *Then (up to functions from* P*)*

$$\psi_1 = C(h)(K_1 + K_2), \tag{6.3}$$

$$\psi_2 = C(h)(K_1 - K_2), \tag{6.4}$$

$$C(h) = (h\pi)^{-n/4}(\omega_1 \cdots \omega_n)^{1/4}, \tag{6.5}$$

where $\omega_j = \omega_j^1 = \omega_j^2$.

PROOF. If $E_1 = E_2$, then the whole space generated by K_1 and K_2 is the eigenspace for $\widehat{H}$ with the eigenvalue $E_1 = E_2$ and our proposition is trivial. We shall suppose further that E_1 and E_2 do not coincide. Let $\psi(x)$ be any eigenfunction of $\widehat{H}$. Then $\psi(R(x))$ is also an eigenfunction with the same eigenvalue. But both low eigenspaces of $\widehat{H}$ are one-dimensional by assumption. It follows that

$$\psi_1(R(x)) = \alpha_1 \psi_1(x), \qquad \psi_2(R(x)) = \alpha_2 \psi_2(x),$$

where α_1, α_2 are certain complex constants from the unit circle. Taking into consideration the fact that $K_1(R(x)) = K_2(x)$ (which is clear from the construction of K_1 and K_2) we get $\alpha_{1,2} = \pm 1$ and (6.3), (6.4) hold. The formula for $C(h)$ follows from the normalization condition for $\|\psi_j\|$ after the calculation of the norms of the functions K_1, K_2 by the Laplace method.

PROPOSITION 2. *Let* $V(x)$ *have three minimum points* ξ_1, ξ_2, *and* ξ_3 *which belong to the plane* $\{x_3 = x_4 = \cdots = x_n = 0\}$ *and moreover let* $V(x)$ *be invariant with respect to the rotation* R *of this plane by the angle* $2\pi/3$, *i.e.,* $V(R(x)) = V(x)$, *which implies, in particular, that* $\xi_2 = R(\xi_1)$, $\xi_3 = R(\xi_2)$. *Then the one-dimensional eigenspace of the lowest energy level* E_1 *is generated by the function*

$$\psi_1 = C(h)(K_1 + K_2 + K_3), \tag{6.6}$$

and the two other levels E_2, E_3 *coincide: their eigenspace is generated by the functions* $K_i - K_j$, $i, j = 1, 2, 3$, *and consists of the functions* $\psi = C_1(h)K_1 + C_2(h)K_2 + C_3(h)K_3$ *such that* $C_1(h) + C_2(h) + C_3(h) = 0$.

PROOF. As in the previous proposition, it is sufficient to consider the case when among the levels E_1, E_2, E_3 there are different ones. Then evidently there exists a one-dimensional eigenspace. As in Proposition 1, using the invariance property of the potential V, we conclude that the eigenfunction ψ_1 of this one-dimensional eigenspace satisfies the following condition: $\psi_1(R(x)) = \alpha\psi_1(x)$ for some α such that $|\alpha| = 1$. Since the transformation R generates the third order transformation group we have $\alpha^3 = 1$. Taking into consideration the fact that ψ_1 can be assumed real and that the transformation R takes real-valued functions to real-valued ones we get $\alpha = 1$, which implies (6.6).

Now let us consider the two-dimensional eigenspace. It is generated by the pair of functions $\psi(x)$ and $\psi'(x) = \psi(R(x))$. Therefore $\psi''(x) = \psi'(R(x))$ is their linear combination: $\psi'' = \alpha\psi + \beta\psi'$. Substituting here $R(x)$ for x we get: $\psi = \alpha\psi' + \beta\psi''$. Substituting it in the previous formula, we obtain $(1-\alpha\beta)\psi = (\alpha+\beta^2)\psi'$. This implies $1-\alpha\beta = 0$ and $\alpha+\beta^2 = 0$. Hence $\beta^3 = -1$. Using again the fact that the eigenfunctions are real-valued we conclude that $\beta = -1$ and, consequently, $\alpha = -1$. Therefore $\psi'' = -(\psi+\psi')$. For functions of the type $\psi = C_1(h)K_1+C_2(h)K_2+C_3(h)K_3$ this means exactly that $C_1(h)+C_2(h)+C_3(h) = 0$. The proposition is proved.

Now we shall calculate the distance, exponentially small in h, between the lowest energy levels E_1, E_2 in the case of a potential with two symmetric minimums, i.e., under the conditions of Proposition 1. (The argument and the calculations in the case of the conditions of Proposition 2 are quite similar.) We shall use here our asymptotic formulas for eigenfunctions and a generalization of a formula for the splitting value proposed in [44, 45]. A physical discussion of the problem and of the related ones can be found in [29, 30, 38, 39, 41–44] and many other publications. The one-dimensional case was considered carefully in [33].

Let us multiply equation (6.1) for the function ψ_1 of the form (6.3) by the function ψ_2 of the form (6.4) and then multiply equation (6.1) for ψ_2 by ψ_1. Now we subtract one of the resulting equations from the other. Then we integrate the result over some domain D with smooth boundary. Thus we get

$$\int_D -(h^2/2)(\psi_1\Delta\psi_2 - \psi_2\Delta\psi_1)\,dx = (E_2 - E_1)\int_D \psi_1\psi_2\,dx. \tag{6.7}$$

Using the Green formula to transform the left-hand side we obtain

$$(E_2 - E_1) = -\frac{h^2}{2}\int_{\partial D}\left(\psi_1\frac{\partial\psi_2}{\partial n} - \psi_2\frac{\partial\psi_1}{\partial n}\right)ds\left(\int_D \psi_1\psi_2\,dx\right)^{-1}. \tag{6.8}$$

Here $\partial/\partial n$ stands for the derivative in the direction of the outer normal unit vector n to the boundary ∂D. In this formula we now substitute the domain D^1 introduced in the previous section instead of the arbitrary domain D. In the case considered, the domain D^1 coincides with the semispace $\{x_1 \geq 0\}$ due to the parity of $V(x)$. Substituting the expressions (6.3), (6.4) for ψ_1, ψ_2 in (6.8) and calculating the denominator on the right-hand side of (6.8) by the Laplace method, we get now

$$\begin{aligned}\int_{D_1}\psi_1\psi_2\,dx &= (C(h))^2\int_{D_1}(K_1^2 - K_2^2)\,dx = (C(h))^2(1+O(h))\int_{D_1} R_1^2\,dx \\ &= (1/2)(1+O(h))\|\psi_1\|^2 = 1/2 + O(h),\end{aligned}$$

$$E_2 - E_1 = 2h^2(C(h))^2(1+O(h))\int_{\partial D^1}\left(K_1\frac{\partial K_2}{\partial n} - K_2\frac{\partial K_1}{\partial n}\right)ds. \tag{6.9}$$

To compute this integral by the Laplace method, we assume at first that there exists a unique nondegenerate point y_0 where the function $S_1(x)|_{\partial D^1}$

has its global minimum (here and further S_k, J_k denote the entropy and the Jacobian on the manifolds Ω^k introduced in §4). The integrand in (6.9) in a neighborhood of y_0 is equal to

$$\begin{aligned} J_1^{-1/2} \exp\{-S_1/h\} \frac{\partial}{\partial n}(J_2^{-1/2} \exp\{-S_2/h\}) \\ - J_2^{-1/2} \exp\{-S_2/h\} \frac{\partial}{\partial n}(J_1^{-1/2} \exp\{-S_1/h\}) \\ = \frac{1}{h}(1+O(h))(J_1 J_2)^{-1/2} \left(\frac{\partial S_1}{\partial n} - \frac{\partial S_2}{\partial n}\right) \exp\left\{ - \frac{1}{h}(S_1 + S_2)\right\} \end{aligned}$$

and this is equal to

$$\frac{2}{h} J_1^{-1} \exp\left\{ - \frac{2}{h} S_1 \right\} \frac{\partial S_1}{\partial n}(1 + O(h))$$

for $S_1(R(x)) = S_2(x)$, $J_1(R(x)) = J_2(x)$ on ∂D^1 due to the parity of the function $V(x)$. Therefore, calculating the integral in (6.9) by the Laplace method, we get

$$\begin{aligned} E_2 - E_1 = 4h(C(h))^2(1+O(h))(\psi h)^{(n-1)/2} \\ \times J_1^{-1}(y_0) \frac{\partial S_1}{\partial n}(y_0) \exp\left\{ - \frac{2}{h} S_1(y_0)\right\} (\det A)^{-1/2}, \end{aligned}$$

where A is the matrix of second derivatives of the function $S_1|_{\partial D^1}$ with respect to the arguments $x_2, \dots, x_n$. Now let $q_1(\tau)$ be the trajectory of the Newton system $\ddot{q} = \partial V/\partial q$ from Proposition 2 of §4, i.e., the trajectory such that $q_1(-\infty) = \xi_1$, $q_1(0) = y_0$, and let the entropy along $q_1(\tau)$ be equal to $S_1(y_0)$. This trajectory is orthogonal to ∂D^1 since y_0 is an extremum of $S_1|_{\partial D^1}$. As $V(x)$ is even, we conclude that the trajectory $q_2(\tau) = Rq_1(\tau)$ is also orthogonal to ∂D^1 and $q_2(-\infty) = \xi_2$, $q_2(0) = y$. It follows that there exists a trajectory $q(\tau)$ of the Newton system such that

$$\begin{gathered} q(-\infty) = \xi_1, \quad q(+\infty) = \xi_2, \quad q(0) = y, \quad \dot{q}(-\infty) = \dot{q}(+\infty) = 0, \\ S_1(y) = S_2(y) = \frac{1}{2}\int_{-\infty}^{+\infty} \dot{q}^2 \, d\tau. \end{gathered} \tag{6.10}$$

A solution of the Newton system $\ddot{q} = V'(q)$ is called an *instanton* of the quantum mechanics problem with the Hamiltonian $H = \dot{q}^2/2 + V'(q)$ if it has the following property

$$\dot{q}(-\infty) = \dot{q}(+\infty) = 0, \qquad q(\mp\infty) = \xi_{1,2}.$$

We denote by S_{12} the minimum of the integral $\int_{-\infty}^{+\infty} \dot{q}^2 \, d\tau$, defined for all instantons. Substituting formula (6.5) for $C(h)$ in (6.9), we get

$$\begin{aligned} E_2 - E_1 = 4\pi^{-1/2} h^{1/2}(1+O(h))(\omega_1 \cdots \omega_n)^{1/2} \\ \times J_1^{-1}(y_0) \frac{\partial S_1}{\partial n}(y_0) \det A^{-1/2} \exp\{-S_{12}/h\}. \end{aligned} \tag{6.11}$$

Note that the derivative $(\partial S_1/\partial n)(y_0)$ in this formula is the modulus of the instanton's velocity $\dot{q}(0)$ at the point y_0.

In particular, from (6.11) we get the formula for the first term of the logarithmic asymptotics of the splitting value $E_2 - E_1$

$$\lim_{h\to 0} h \log(E_2 - E_1) = -S_{12}. \tag{6.12}$$

If there exist several nondegenerate points where the function $S_1|_{\partial D^1}$ has a global minimum, then $E_2 - E_1$ is equal to the sum of the expressions on the right-hand side of (6.11) corresponding to all such points. If the minimum of the function $S_1|_{\partial D^1}$ is degenerate, then formula (6.11) does not hold any more. But in this situation one can use the result from [8] to calculate the first term of the logarithmic asymptotics of the value $E_2 - E_1$. Therefore, formula (6.12) remains true in this more general case.

Similarly one can get asymptotic formulas for low energy level splitting in the case of the potential described in Proposition 2 and also for other potentials with discrete symmetry.

Now let us show how to calculate the preexponential term in (6.11). The matrix A there is equal to

$$\left\| \frac{\partial^2 S}{\partial x_j \partial x_k}(y_0) \right\|_{k,j=2,3,\dots,n}.$$

Let us consider the system in variations along the instanton $(q(t), p(t))$

$$\dot{Z} = W, \qquad \dot{W} = V_{xx}(q(t))Z \tag{6.13}$$

or the system

$$\ddot{Z} = V_{xx}(q(t))Z \tag{6.14}$$

for Z. We consider the matrix-valued solution of this system with the boundary conditions

$$Z|_{t=0} = I, \qquad Z|_{t\to-\infty} = 0, \tag{6.15}$$

where I is the unit matrix.

PROPOSITION 3. A) *There exists a unique solution $Z(t)$ of the problem* (6.14), (6.15);

B) $(\partial^2 S/\partial x^2)(y_0) = \dot{Z}(0)$;

C) $J_1^{-1}(y_0) = \lim_{t\to-\infty}[\exp\{-(\omega_1 + \dots + \omega_n)t\} \det Z(t)]$.

PROOF. For simplicity we restrict ourselves to the case of a nonresonance ω_j. In the coordinates $\mu = (\mu_1, \dots, \mu_n)$ (see Proposition 4) the manifold W_k^{out} is given by the equation $q = f(\mu)$, $p = g(\mu)$, where f, g are certain vector-functions. The phase flow g_H^t (2.3) takes the point (x, p) with coordinate μ to the point

$$(q(t), p(t)) = (f, g)(e^{\omega_1 t}\mu_1, \dots, e^{\omega_n t}\mu_n).$$

Let μ^0 be the point such that $q = f(\mu^0) = y_0$. Then

$$J_1(y_0) = \det(\partial f(\mu)/\partial \mu)$$

and the equations

$$q = f(e^{\omega_1 t}\mu_1^0, \dots, e^{\omega_n t}\mu_n^0), \qquad p = g(e^{\omega_1 t}\mu_1^0, \dots, e^{\omega_n t}\mu_n^0)$$

define the instanton. Differentiating f and g in μ, we get the solution $a_j(t) = (\widetilde{W}_j(t), \widetilde{Z}_j(t))$ of the system (6.13), where

$$\widetilde{Z}_j = \left(\frac{\partial}{\partial \mu_j} f(e^{\omega_1 t}\mu_1, \dots, e^{\omega_n t}\mu_n)\right)\Big|_{\mu=\mu^0},$$
$$\widetilde{W}_j = \left(\frac{\partial}{\partial \mu_j} g(e^{\omega_1 t}\mu_1, \dots, e^{\omega_n t}\mu_n)\right)\Big|_{\mu=\mu^0}.$$

Note that $a_1(t), \dots, a_n(t)$ are independent, since

$$J_1 = \det \frac{\partial f}{\partial \mu}(\mu^0) \neq 0,$$

and $a_j \to 0$ as $t \to -\infty$. Therefore, the vector-functions $(\widetilde{Z}_1(t), \dots, \widetilde{Z}_n(t))$ determine a basis in the space of the solutions of system (6.13) that are bounded as $t \to -\infty$. It follows that the solution $Z(t)$ of problem (6.14), (6.15) is equal to $(\widetilde{Z}_1(t), \dots, \widetilde{Z}_n(t))(\widetilde{Z}_1(0), \dots, \widetilde{Z}_n(0))^{-1}$. This proves A). From the last formula we also get

$$J_1 = \det(\widetilde{Z}_1(t), \dots, \widetilde{Z}_n(t))/\det Z(t)$$

for all t. Now we get C) using (5.13) and (5.22). Assertion B) follows from the evident formula

$$\frac{\partial^2 S}{\partial x^2} = \frac{\partial g}{\partial \mu}\left(\frac{\partial f}{\partial \mu}\right)^{-1}(\mu^0).$$

Thus formula (6.11) takes the form

$$\begin{aligned} E_2 - E_1 &= 4\pi^{-1/2} h^{1/2} (\omega_1 \cdots \omega_n)^{1/2} |\dot{q}(0)| (1 + O(h)) \\ &\times \det \left\| \frac{\partial^2 S}{\partial x_j \partial x_k}(y_0) \right\|_{k,\, j=2,3,\dots,n} \exp\left\{ -\frac{1}{2}\int_{-\infty}^{\infty} \dot{q}^2(t)\, dt \right\} \\ &\times \lim_{t\to-\infty} [\exp\{-(\omega_1 + \cdots + \omega_n)t\} \det Z(t)]. \end{aligned} \tag{6.16}$$

REMARK. The formula for the amplitude of the splitting value in (6.6) contains only the matrix of the second derivatives of the phase calculated on the instanton. It means that only a Gaussian packet (or a sqeezed state) stands for the tunneling. Its size defines a tube of tunneling in a neighborhood of the instanton.

§7. The tunnel operator on the torus, and the Schrödinger equation with a periodic potential

In this section we describe briefly the theory of the periodic tunnel operator and its application to the construction of exponential asymptotics of low lying eigenfunctions of the periodic Schrödinger operator. The theory of periodic Schrödinger operators is largely used in the theory of solids.

Let $V(x)$ be 2π-periodic in each coordinate. We consider the heat conduction equation

$$h\frac{\partial u}{\partial t} = \frac{h^2}{2}\Delta u - V(x)u \tag{7.1}$$

in the class of 2π-periodic (in each coordinate) functions on $\mathbb{R}^n_x$. In other words, we regard $u(x)$ as a function on the n-dimensional torus $T^n = \mathbb{R}^n/(2\pi\mathbb{Z})^n$. The solution $u^p(t, x, \xi, h)$ periodic in x of the equation (7.1) is said to be the *Green function* of the periodic Cauchy problem for equation (7.1) if it satisfies the initial condition

$$u^p(0, x, \xi, h) = \sum_{N\in\mathbb{Z}^n} \delta(x - (\xi + 2\pi N)). \tag{7.2}$$

We assume the smooth function $V(x)$ to be nonnegative and to have only finite number of zero points $\xi_1, \dots, \xi_l$ on the torus T^n. Evidently we have

$$u^p(t, x, \xi, h) = \sum_{N\in\mathbb{Z}^n} u(t, x, \xi + 2\pi N, h), \tag{7.3}$$

where $u(t, x, \xi, h)$ is the usual (nonperiodic) Green function of the form (2.5) for small t and of the form (4.1) or (4.2) for all finite t. Since the entropy $S(t, x, \xi)$ tends to infinity as $\xi \to \infty$, it follows that the series in (7.3) converges and that only a finite number of terms actually contribute to the asymptotics.

We shall assume further that the matrices of second derivatives $V''(\xi_k)$ of the potential V at the zero points $\xi_1, \dots, \xi_l$ are nondegenerate and that the value $\mathscr{E} = \sum_{j=1}^n \omega_j^k$ does not depend on k, where $(\omega_j^k)^2$, $j = 1, \dots, n$, are the eigenvalues of the matrix $V''(\xi_k)$. We shall also denote ξ_k the inverse image of $\xi_k \in T^n$ in the square $[0, 2\pi]^n \in \mathbb{R}^n$.

Let us find the asymptotics of the solution u_k^p of the periodic Cauchy problem with initial function

$$u_k^p(0, x, h) = \sum_{N\in\mathbb{Z}^n} \chi_k(x + 2\pi N), \tag{7.4}$$

where χ_k is the characteristic function of a small closed neighborhood V^k of the point $\xi_k \in \mathbb{R}^n$. For this purpose we introduce the tunnel operators $K_{k,N}$, defined by a countable number of zero points

$$\xi_{k,N} = \xi_k + 2\pi N, \qquad k = 1, \dots, l, \quad N \in \mathbb{Z}^n,$$

of the potential $V(x)$ in $\mathbb{R}^n$ as described in §5 (see the notation after formula (5.15)). In particular, the domain $D^{k,N}$, which is involved in the construction of the operator, consists of points x such that

$$S_{k,N}(x) = \min_{j,M}\{S_{j,M}(x)\}. \tag{7.5}$$

By the superposition principle and Theorem 6, we have

$$u_k^p = \exp\{-(\mathscr{E}/2)h^{-(1+\varkappa)}\} \sum_{N\in\mathbb{Z}^n} K_{k,N}(1+O(h)) \tag{7.6}$$

as $h\to 0$, $t = h^{-(1+\varkappa)}$, $\varkappa \in (0, 1)$.

Now let us consider the Schrödinger equation on the torus

$$\widehat{H}\psi = E\psi, \qquad \widehat{H} = -(h^2/2)\Delta + V(x), \quad \psi \in L_2(T^n). \tag{7.7}$$

In other words, we look for periodic solutions of equation (7.7) in $\mathbb{R}^n$. It is well known that this problem has positive discrete spectrum. As in the case of the Schrödinger operator considered in the previous section, the operator $\widehat{H}$ on the torus has precisely l eigenvalues $E_1, \dots, E_l$ of the form $h(\mathscr{E}/2+O(h))$ and the inequality $(E/h)-(\mathscr{E}/2) \geqslant C > 0$ holds for all other eigenvalues E (see, for instance, [22]). As in Theorem 7, one can show that the low eigenfunctions $\psi_1, \dots, \psi_l$ prove to be linear combinations

$$\psi_j = \sum_{i=1}^{l} C_i(h)\widetilde{K}_i$$

of the periodic tunnel canonical operators

$$\widetilde{K}_j = \sum_{N\in\mathbb{Z}^n} K_{j,N}.$$

Let $V(x)$ have only two minimum points in T^n that are images of the points

$$\xi_1 = (|\xi_1|, 0, \dots, 0), \qquad \xi_2 = (-|\xi_1|, 0, \dots, 0),$$

where $|\xi_1| < \pi$, and let V be even in each coordinate. Then it is clear that the domains $D^{k,N} = D^{k,0} + N$, $k = 1, 2$, are parallelepipeds:

$$D^1 = \{x \in \mathbb{R}^n : x_1 \in [0, \pi],\ \forall j > 1,\ x_j \in [-\pi, \pi]\},$$
$$D^2 = \{x \in \mathbb{R}^n : x_1 \in [-\pi, 0],\ \forall j > 1,\ x_j \in [-\pi, \pi]\}.$$

As in Proposition 1 of the previous section, we get $\psi_{1,2} = C(h)(\widetilde{K}_1 \pm \widetilde{K}_2)$.

The specificity of the periodic case becomes more significant when one calculates the splitting value $E_2 - E_1$. Now the limit

$$-\lim_{h\to 0} h\log(E_2 - E_1)$$

is equal to the minimum of the actions $(1/2)\int \dot{q}^2 d\tau$ along all instantons joining ξ_1 to ξ_2 as well as ξ_1 to ξ_1. In a similar way one must modify formula (6.11) in the nondegenerate situation.

It is well known (see, for example, [13, 14]) that the investigation of the spectrum of a Schrödinger operator with periodic potential $V(x)$ on the space $L^2(\mathbb{R}^n)$ (without the periodicity condition for the solution) may be reduced in a certain sense to the solution of some auxiliary problems on the torus. Let us illustrate this thesis, using our results in the investigation of the one-dimensional Schrödinger equation

$$-(h^2/2)\psi''(x) + V(x)\psi(x) = E\psi(x) \tag{7.8}$$

on the line $x \in \mathbb{R}$ with nonnegative even 2π-periodic potential V having a single zero point $\xi = \pi$ on the segment $[0, 2\pi]$. The spectrum of this problem is known to be absolutely continuous and to consist of a countable number of segments $[\alpha_1, \alpha_2] \cup [\alpha_3, \alpha_4] \cup \cdots$. Here α_{2j-1} (respectively α_{2j}) is the jth eigenvalue of the operator

$$\widehat{H}\psi(x) = -(h^2/2)\psi''(x) + V(x)\psi(x)$$

on the line with the periodicity condition $\psi(x + 2\pi) = \psi(x)$ (respectively with the antiperiodicity condition $\psi(x + 2\pi) = -\psi(x)$). These facts can be found, for example, in [13]. It follows evidently that α_1, α_2 are the low levels of the operator $\widehat{H}$ defined on the duplicated circle. Therefore the calculation of the width of the first allowed zone $[\alpha_1, \alpha_2]$ of the spectrum of equation (7.8) is reduced to the calculation of the splitting of low lying eigenvalues for the operator on the circle with nonnegative potential having two zero points. The solution of this problem is described above. Thus

$$\lim_{h\to 0} h \log(\alpha_2 - \alpha_1) = -S_{12},$$

where S_{12} is the action along an instanton. This formula is well known in the physical literature [23, 31, 37].

Using our methods, one can calculate the first term of the exponential asymptotics of the splitting value $\alpha_2 - \alpha_1$. The nondegeneracy condition which one needs to use formula (6.11) is satisfied automatically in the one-dimensional situation. Using the segment $[0, 2\pi]$ instead of the general domain D in the one-dimensional periodic version of formula (6.9), we get

$$\alpha_2 - \alpha_1 = 2h^2(C(h))^2(1 + O(h))(\widetilde{K}_1\widetilde{K}_2' - \widetilde{K}_2\widetilde{K}_1')\big|_0^{2\pi}$$

which implies

$$\alpha_2 - \alpha_1 = 8(hV''(\pi)/\pi)^{1/2} J_1^{-1}(0)(1 + O(h))\dot{y}(0)\exp\{-(2/h)S_1(0)\}, \tag{7.9}$$

where $\dot{y}$ is the velocity of the instanton joining $-\pi$ and π.

§8. Example: a chain on the circle with pairwise interaction

Let us consider the Schrödinger equation (6.1) with potential $V(x)$, having two symmetric minimum points:

$$V = \frac{\alpha}{2}\sum_{k=1}^{n}(x_k - x_{k-1})^2 + \frac{\beta}{4}\sum_{k=1}^{n}(x_k - \xi)^2(x_k + \xi)^2.$$

Here α, β, ξ are positive parameters and $x_0 = x_n$ by definition. It is clear that $V(x) \geqslant 0$, $V(x) \to \infty$ as $|x| \to \infty$, $V(x) = V(-x)$ for all x, and $V(x)$ has only two global minimum points $x^{\pm} = \pm(\xi, \xi, \dots, \xi)$. Let us find the eigenvalues ω_m^2 of the matrix $(\partial^2 V/\partial x_i \partial x_j)$. They satisfy the equations

$$-\alpha(z_{k+1} - 2z_k + z_{k-1}) + 2\beta\xi^2 z_k = \omega^2 z_k, \qquad k = 1, \dots, n, \ z_0 = z_n.$$

Their solutions are well known [36]:

$$\omega_m^2 = 4\alpha \sin^2 \frac{\pi m}{n} + 2\beta\xi^2, \qquad (z_m)_k = \frac{1}{\sqrt{h}} \exp\left\{ i\frac{2\pi}{h} mk \right\},$$

where $m = 0, 1, \dots, n-1$. Thus the formula for the low energy levels has the form

$$E_0^{\pm} = \frac{h}{2} \sum_{m=0}^{n-1} \sqrt{4\alpha \sin^2 \frac{\pi m}{n} + 2\beta\xi^2} + O(h^2).$$

Let us calculate the splitting $\Delta = E^+ - E^-$. First we find the instanton solution of the Newton system:

$$\ddot{x}_k = \beta(x_k^3 - x_k\xi) - \alpha(x_{k+1} - 2x_k + x_{k-1}),$$
$$x_k\big|_{t=-\infty} = -\xi, \quad x_k\big|_{t=\infty} = \xi, \qquad k = 1, \dots, n.$$

One of solutions has the form $x_1 = x_2 = \cdots = x_n = \sqrt{\xi} \tanh(t\sqrt{\beta\xi/2})$. Let us show that this solution gives the minimum of the integral

$$\int_{-\infty}^{\infty} \left(\frac{\dot{x}^2}{2} + V(x(t)) \right) dt$$

among all continuous piecewise-smooth curves with endpoints $x(-\infty) = -\xi$, $x(\infty) = \xi$. In fact,

$$\begin{aligned} 2S &= \min \sum_{k=1}^{n} \int_{-\infty}^{\infty} \left[\frac{\dot{x}_k^2}{2} + \frac{\alpha}{2}(x_k - x_{k-1})^2 + \frac{\beta}{4}(x_k + \xi)^2 (x_k - \xi)^2 \right] dt \\ &\geqslant \min \sum_{k=1}^{n} \int_{-\infty}^{\infty} \left[\frac{\dot{x}_k^2}{2} + \frac{\beta}{4}(x_k + \xi)^2 (x_k - \xi)^2 \right] dt \\ &= n \min_{q(t)\,:\, q(\pm\infty) = \pm\xi} \int_{-\infty}^{\infty} \left[\frac{\dot{q}^2}{2} + \frac{\beta}{4}(q + \xi)^2 (q - \xi)^2 \right] dt. \end{aligned}$$

The solution $q = \sqrt{\xi} \tanh(\sqrt{\beta}\xi t/\sqrt{2})$ of the system $\ddot{q} = \beta(q^3 - \xi q)$ gives the minimum of the last expression.

Now let us calculate $2S$. Since $\dot{q}^2 = \beta(q - \xi)^2 (q + \xi)^2/4$, we have

$$2S = n \int_{-\infty}^{\infty} \dot{q}^2 \, dt = \frac{\sqrt{\beta} h}{2} \int_{-\xi}^{\xi} (\xi - x)(\xi + x) \, dx = \frac{2}{3} n \sqrt{\beta} \xi^2.$$

Therefore,

$$E_1 - E_2 = A \exp\{-2n\sqrt{\beta}\xi^3/(3h)\}. \tag{8.1}$$

This formula can be generalized to the Schrödinger equation with the potential

$$V = \sum_{k=1}^{n} \alpha_k (x_k - x_{k-1})^2 + \frac{\beta}{4} \sum_{k=1}^{n} u(x_k),$$

where $\alpha_k > 0$ and $u(z)$ is an even function that has two nondegenerate global minimum points and tends to infinity as $|x| \to \infty$. Using formula (6.16), we also can easily calculate the coefficient A in (8.1).

§9. Hydrogen ion in a magnetic field

As another example we consider the Schrödinger equation for the hydrogen ion. First of all we note that the theory of the previous sections can be applied to more general potentials. Namely, the potential V may have several isolated singular points and need not tend to infinity as $|x| \to \infty$.

We start with a problem that can be solved exactly by separating the variables. This problem was discussed in many papers. In particular, in [34] the quasiclassical asymptotics were obtained for high energy levels. Here we get series of eigenvalues and their splitting that cannot be obtained by formally applying the method of [34]. A physical discussion of the problem can be found in [22, 39, 40]. The potential has the form

$$V = -\left(\frac{\alpha}{\sqrt{x^2 + y^2 + (z-a)^2}} + \frac{\alpha}{\sqrt{x^2 + y^2 + (z+a)^2}} \right).$$

Here α, a are physical constants, $\pm a$ define the points on the z axis where the atoms are placed. Let us write out the problem in cylindrical coordinates ρ, ϕ, z, taking into consideration the angular symmetry and thus representing the wave function in the form $e^{im\phi}\psi(\rho, z)$ (m is an integer):

$$-\frac{h^2}{2}\left(\frac{1}{\rho}\frac{\partial}{\partial \rho}\rho\frac{\partial}{\partial \rho} + \frac{\partial^2}{\partial z^2}\right)\psi + V_1\psi = E\psi,$$

$$V_1(\rho, z) = \frac{p_\phi^2}{2\rho^2} - \alpha\left\{\frac{1}{\sqrt{\rho^2 + (z-a)^2}} + \frac{1}{\sqrt{\rho^2 + (z+a)^2}}\right\},$$

where $p_\phi = mh$. It is evident that V_1 is bounded from below, if $|p_\phi| \geqslant C$ for some $C > 0$. We will show that this potential has two points of global minimum, if $p_\phi^2 \leqslant 8a\alpha/(3\sqrt{3})$. To apply the above theory, it is convenient to use the special (ellipsoidal) coordinates σ, τ, ϕ such that $\sigma \geqslant 1 \geqslant \tau \geqslant -1$, $\phi \in [0, 2\pi]$ and $z = a\sigma\tau$, $\rho^2 = \alpha^2(\sigma^2 - 1)(1 - \tau^2)$. In these coordinates the equation for ψ has the form

$$-\frac{h^2}{\sigma^2 - \tau^2}\left(\frac{\partial}{\partial \sigma}(\sigma^2 - 1)\frac{\partial}{\partial \sigma} + \frac{\partial}{\partial \tau}(1 - \tau^2)\frac{\partial}{\partial \tau}\right)\psi + V_{\text{eff}}\psi = 2a^2 E\psi, \qquad (9.1)$$

where

$$V_{\text{eff}} = \frac{1}{\sigma^2 - \tau^2}\left[p_\phi^2\left(\frac{1}{\sigma^2 - 1} + \frac{1}{1 - \tau^2}\right) - 4a\alpha\sigma\right]$$

and $p_\phi = hm$. Further we consider m to be of order $1/h$ (in fact, the large number m is a parameter in the problem considered). The potential V_{eff} has two points of global minimum $(\tau_\pm, \sigma_0)$ that can be determined in the following way. The number σ_0 is the unique root of the equation

$$\sigma^3 = \frac{a\alpha}{4p_\phi^2}(\sigma^2+1)^2(\sigma^2-1).$$

The value $\mathscr{E} = -V_{\text{eff}}(\tau_\pm, \sigma_0)$ and points $\tau_\pm$ can be defined by the formulas

$$\mathscr{E} = |p_\phi|\frac{1}{(\sigma_0^2-1)^2} + \frac{2\alpha}{p_\phi^2\sigma_0}, \qquad \tau_\pm = \pm\sqrt{1 - \frac{p_\phi}{\sqrt{\mathscr{E}}}}.$$

By putting $\widetilde{E} = 2a^2E + \mathscr{E}$ and $\widetilde{V}_{\text{eff}} = V_{\text{eff}} + \mathscr{E}$, we reduce equation (9.1) to the form

$$-\frac{h^2}{\sigma^2-\tau^2}\left(\frac{\partial}{\partial\sigma}(\sigma^2-1)\frac{\partial}{\partial\sigma} + \frac{\partial}{\partial\tau}(1-\tau^2)\frac{\partial}{\partial\tau}\right)\psi + \widetilde{V}_{\text{eff}}\psi = 2a^2\widetilde{E}\psi, \qquad (9.2)$$

where $\widetilde{V}_{\text{eff}} \geqslant 0$. Now we may use the oscillator approximation. The frequences ω_j can be defined as the eigenvalues of the matrix (see, [5])

$$\mathscr{H} = \begin{pmatrix} -H_{xp} & -H_{xx} \\ H_{pp} & H_{px} \end{pmatrix},$$

whose elements are the second derivatives of H calculated at $\sigma = \sigma_0$, $\tau = \tau_\pm$. Here H is the tunnel Hamiltonian

$$H = (\sigma^2-\tau^2)^{-1}(p_\sigma^2(\sigma^2-1) + p_\tau^2(1-\tau^2)) - \widetilde{V}_{\text{eff}},$$

and $p_1 = p_\sigma$, $p_2 = p_\tau$, $x_1 = \sigma$, $x_2 = \tau$. Straightforward but rather tiresome calculations give

$$\omega_1 = \frac{|p_\phi|(3\sigma_0^2-1)\sqrt{3\sigma_0^4 - 2\sigma_0^2 + 3}}{\sigma_0(\sigma_0^2-1)(\sigma_0^4-1)},$$

$$\omega_2 = \frac{|p_\phi|(3\sigma_0^2-1)\sqrt{(3\sigma_0^2-1)(3-\sigma_0^2)}}{\sigma_0(\sigma_0^2-1)(\sigma_0^4-1)}.$$

Thus the low eigenvalues of problem (9.2) have the form

$$\widetilde{E}_{0,1} = (1/2)(h\omega_1 + h\omega_2) + O(h^2).$$

In the classical limit these eigenvalues correspond to the rotation of the electron on the circle $\sigma = \sigma_0$, $\tau = \tau_\pm$. Using formula (6.11) let us find the splitting value. We have

$$-h\log(\widetilde{E}_1 - \widetilde{E}_0) = \int_{\xi_-}^{\xi_+} p\,dx + O(h),$$

where $\xi_\pm = (\sigma_0, \tau_\pm)$. It is easy to show that $p_\sigma \equiv 0$, $\sigma = \sigma_0$ on the instanton solution. Therefore,

$$\int_{\xi_-}^{\xi_+} p\,dx = \int_{\tau_-}^{\tau_+} p_\tau\,d\tau.$$

Using the law of conservation of energy we get

$$p_\tau = |p_\phi| \frac{\tau_+^2 - \tau^2}{(1-\tau^2)(1-\tau_+^2)}.$$

It follows that

$$\int_{\tau_-}^{\tau_+} p_\tau\,d\tau = 2\tau_+\sqrt{\mathscr{E}} + |p_\phi| \log\frac{1-\tau_+}{1+\tau_+}.$$

Finally we obtain the series of eigenfunctions of the initial problem

$$E_{0,1}^m = -(1/2)a^{-2}(h\omega_1 + h\omega_2) + O(h^2), \tag{9.3}$$

with the splitting values

$$-h\log(E_0^m - E_1^m) = |p_\phi| \log\frac{1-\tau_+}{1+\tau_+} + 2\tau_+\sqrt{\mathscr{E}(|p_\phi|)} + O(h), \tag{9.4}$$

where $p_\phi = mh$.

The techniques developed above can be applied also in the more difficult case of the hydrogen ion in a constant magnetic field with the vector potential of the form $A = (-\beta y, \beta x, 0)$ in orthogonal coordinates, where β is a physical constant. In this case the potential V_{eff} in (9.1) has the form

$$V_{\text{eff}} = \frac{p_\phi^2}{(\sigma^2-1)(1-\tau^2)} + 4\beta a^2 p_\phi + 2a^4\beta^2(\sigma^2-1)(1-\tau^2) - \frac{4a\alpha\sigma}{\sigma^2-\tau^2},$$

and as in the previous case, it has two points of global minimum $(\sigma_0, \tau_\pm)$, if

$$|p_\phi| \leqslant 8a\alpha/(3\sqrt{3}) + 8a^4\beta^2,$$

where $\sigma_0 = \sigma_0(p_\phi)$ is the unique positive root of the equation

$$4p_\phi^2\sigma^3(3\sigma^2-1)^2 = a\alpha(3\sigma^2-1)^2(\sigma^2+1)^2(\sigma^2-1) + 8a^4\beta^2(\sigma^2+1)^2(\sigma^2-1)^4\sigma^3,$$

and the coordinates $\tau_\pm = \tau_\pm(p_\phi)$ are defined by the formula

$$\tau_\pm = \pm\sqrt{\frac{3\sigma_0^2 - \sigma_0^4}{3\sigma_0^2 - 1}}.$$

The eigenvalues $E_{0,1}$ are defined by the same formula (9.3), but now the explicit expression for the frequences $\omega_{0,1}$ is rather cumbersome and we omit it. The value of the exponential splitting of the energy levels is given by the formula

$$
\begin{aligned}
-h\log(E_0^m - E_1^m) = 2\sqrt{2}a^2\beta\sqrt{\sigma_0^2-1} \\
\times\Bigg(\frac{1}{2}\tau_+\sqrt{b^2-\tau_+^2} + \left(\frac{b^2}{2}+2(1-\tau_+^2)\right)\arcsin\frac{\tau_+}{b} \\
+(1-\tau_+^2)\sqrt{b^2-1} \\
\times\log\left|\frac{(1+\tau_+)(\sqrt{(b^2-1)(b^2-\tau_+^2)}+b^2-\tau_+)}{(1-\tau_+)(\sqrt{(b^2-1)(b^2-\tau_+^2)}+b^2+\tau_+)}\right|\Bigg),
\end{aligned}
$$

where

$$b^2 = \frac{m^2h^2}{2a^4\beta^2(\sigma_0^2-1)(1-\tau_+^2)} - 1.$$

REFERENCES

1. V. P. Maslov, *Global exponential asymptotic behavior of the solutions of equations of tunnel type*, Dokl. Akad. Nauk SSSR **261** (1981), no. 5, 1058–1062; English transl. in Soviet Math. Dokl. **24** (1981).

2. ———, *Global exponential asymptotics of the solutions of the tunnel equations and the large deviation problems*, Trudy Mat. Inst. Steklov. **163** (1984), 150–180; English transl. in Proc. Steklov Inst. Math. **1985**, no. 4.

3. ———, *Perturbation theory and asymptotic methods*, Izdat. Moskov. Univ., Moscow, 1965; French transl., Dunod, Paris, 1972.

4. ———, *Asymptotic methods and perturbation theory*, "Nauka", Moscow, 1988. (Russian)

5. ———, *The complex WKB-method in nonlinear equations*, "Nauka", Moscow, 1981. (Russian)

6. ———, *Quasiclassical approximation: asymptotics of some problems of mathematical physics*, Zh. Vychisl. Mat. i Mat. Fiz. **1** (1961), no. 4, 638–663. (Russian)

7. V. P. Maslov and M. V. Fedoryuk, *Semi-classical approximation in quantum mechanics*, "Nauka", Moscow, 1976; English transl., Reidel, Dordrecht, 1981.

8. ———, *Logarithmic asymptotics of the Laplace integrals*, Mat. Zametki **30** (1981), no. 5, 763–768; English transl. in Math. Notes **30** (1981).

9. V. N. Kolokoltsov and V. P. Maslov, *Idempotent calculus as the apparatus of optimization theory*, I, Funktsional. Anal. i Prilozhen. **23** (1989), no. 1, 1–14; II, **23** (1989), no. 4, 53–62; English transl. in Functional Anal. Appl. **23** (1989).

10. M. V. Fedoryuk, *Asymptotics, integrals and sums*, "Nauka", Moscow, 1987. (Russian)

11. F. Hartman, *Ordinary differential equations*, Wiley, New York, 1964.

12. A. A. Borovkov, *Boundary values for random walks and large deviations in functional spaces*, Teor. Veroyatnost. i Primenen. **12** (1967), no. 4, 635–654; English transl. in Theory Probab. Appl. **12** (1967).

13. E. Titchmarsh, *Eigenfunction expansions associated with second order ordinary differential equations*, Part 2, Oxford Univ. Press, Oxford, 1962.

14. M. Reed and B. Simon, *Methods of modern mathematical physics*. Vol. 4, *Analysis of operators*, Academic Press, New York, 1978.

15. J. Palis and W. de Melo, *Geometric theory of dynamical systems. An introduction*, Springer-Verlag, New York, Heidelberg, and Berlin, 1982.

16. V. P. Maslov, *Asymptotic methods for solving pseudodifferential equations*, "Nauka", Moscow, 1987. (Russian)

17. S. R. Varadhan, *Asymptotic probabilities and differential equations*, Comm. Pure Appl. Math. **19** (1966), no. 3, 231–281.

18. B. Simon, *Semiclassical analysis of low lying eigenvalues*. Part 1, *Nondegenerate minima: Asymptotic expansions*, Ann. Inst. H. Poincaré **38** (1983), 295–307.

19. ———, *Semiclassical analysis of low lying eigenvalues*. Part 2, *Tunneling*, Ann. of Math. **120** (1984), 89–118.

20. E. Witten, *Supersymmetry and Morse theory*, J. Differential Geom. **17** (1982), 661–692.

21. A. M. Polyakov, *Quark confinement and topology of gauge theories*, Nuclear Phys. B **120** (1977), 429–456.

22. Martin G. Gutzwiller, *Chaos in classical and quantum mechanics*, Springer-Verlag, New York, 1990.

23. R. Rajaraman, *Solitons and instantons*, North-Holand, Amsterdam, New York, and Oxford, 1982.

24. F. A. Berezin and M. A. Shubin, *The Schrödinger equation*, Izdat. Moskov. Univ., Moscow, 1983; English transl., Reidel, Dordrecht (to appear).

25. S. Yu. Dobrokhotov, V. N. Kolokoltsov, and V. P. Maslov, *The splitting of the low lying energy levels of the Schrödinger operator and the asymptotics of the fundamental solution of the equation* $hu_t = (h^2\Delta/2 - V(x))u$, Teoret. Mat. Fiz. **87** (1991), no. 3, 323–375; English transl. in Theoret. and Math. Phys. **87** (1991).

26. G. Jona-Lasinio, F. Martinelli, and E. Scoppola, *New approach to the semiclassical limit of quantum mechanics*, 1. *Multiple tunneling in one dimension*, Comm. Math. Phys. **80** (1981), 223–254.

27. J. M. Combes, P. Duclos, and R. Seiler, *Convergent expansion for tunneling*, Comm. Math. Phys. **92** (1983), 229–245.

28. S. Sternberg, *Local contractions and a theorem of Poincaré*, Amer. J. Math. **79** (1957), 809–824.

29. E. Gildener and A. Patrascigin, *Pseudoparticle contributions to the energy spectrum of a one-dimensional system*, Phys. Rev. D **16** (1977), 425–443.

30. E. M. Harrel, *Double wells*, Comm. Math. Phys. **75** (1980), 239–261.

31. ———, *The band structure of a one-dimensional periodic system in a scaling limit*, Ann. Phys. **119** (1979), 351–369.

32. B. Heffler and J. Sjöstrand, *Multiple wells in the semiclassical limit*, Ann. Inst. H. Poincaré **46** (1987), no. 4, 353–372.

33. T. F. Pankratova, *Quasimodes and the splitting of eigenvalues*, Dokl. Akad. Nauk SSSR **276** (1984), no. 4, 795-798; English transl. in Soviet Math. Dokl. **29** (1984).

34. V. M. Babich, *Asymptotic solution of the problem about a point source of diffusion in a moving media*, Dokl. Akad. Nauk SSSR **289** (1986), no. 4, 836–839; English transl. in Soviet Math. Dokl. **34** (1986).

35. H. L. Cycon, R. G. Froese, W. Kirsch, and B. Simon, *Schrödinger operator with applications to quantum mechanics and global geometry*, Springer-Verlag, New York, Heidelberg, and Berlin, 1987.

36. O. Madelung, *Festkörpertheorie*, Springer-Verlag, Berlin, Heidelberg, and New York, 1972.

37. S. Coleman, *The uses of instantons*, The Whys of Subnuclear Physics, New York, 1979, pp. 805–916.

38. A. Auerbach and S. Kivelson, *The path decomposition expansion and multidimensional tunneling*, Nuclear Phys. B **275** (1985), 799–858.

39. M. Wilkinson and J. H. Hannay, *Multidimensional tunneling between excited states*, Phys. D **27** (1987), 201–212.

40. M. Wilkinson, *Tunneling between tori in phase space*, Phys. D **21** (1986), 341–354.

41. Z. H. Huang, T. E. Feuchtwang, P. H. Cutler, and E. Kazes, *Wentzel-Kramers-Brillouin method in multidimensional tunneling*, Phys. Rev. A **41** (1990), no. 1, 32–41.

42. J. Callan and S. Coleman, *Fate of the false vacuum,* 2, Phys. Rev. D **16** (1977), no. 6, 1762–1768.

43. P. H. Frampton, *Consequences of vacuum instability in quantum field theory*, Phys. Rev. D **15** (1977), no. 10, 2922–2928.

44. S. Coleman, *Fate of the false vacuum*: *Semiclassical theory*, Phys. Rev. D **15** (1977), no. 10, 2929–2936.

45. L. D. Landau and E. M. Lifshits, *Quantum mechanics*, Moscow, 1963; English transl., Addison-Wesley, Reading, Ma., 1965.

46. C. Herring, *Critique of the Heitler-London method of calculating spin couplings at large distances*, Rev. Modern Phys. **34** (1962), no. 4, 631–645.

Institute for Problems in Mechanics, Russian Academy of Sciences, pr. Vernadskogo 101, 117526, Moscow, Russia

Center for Optimization and Mathematical Modeling, Institute of New Technologies, ul. Kirovogradskaya, 11, 113587, Moscow, Russia

Moscow Institute of Electronic Engineering, B. Vuzovskiĭ per., 3/12, 109028, Moscow, Russia

Translated by V. N. KOLOKOLTSOV

ADVANCES IN SOVIET MATHEMATICS
Volume 13, 1992

Endomorphisms of the Semimodule of Bounded Functions

P. I. DUDNIKOV

§1. Introduction

In the present paper some spectral properties of endomorphisms acting in the semimodule of bounded functions with values in a semiring K are investigated. The semiring K is assumed to satisfy the idempotence, stabilization, and other conditions that are natural for idempotent analysis. The mapping

$$(\Lambda f)(x) = \sup_y K(x, y) \odot f(y),$$

which arises in optimization problems, is an example of such an endomorphism. An upper estimate for eigenvalues of the endomorphisms is obtained in the paper. This estimate is analogous to a corresponding estimate of the spectrum of a linear bounded operator via the spectral radius in ordinary linear algebra.

But, unlike the classical case, in idempotent analysis the availability of two-sided estimates for the spectrum is a typical situation [1]. A corresponding assertion is proved for so-called indecomposable endomorphisms. (The indecomposability concept is similar to the corresponding one in the Frobenius-Perron nonnegative matrix theory.)

Finally, the class of endomorphisms having at most one eigenvalue is described, and some sufficient conditions which guarantee the existence of eigenvalues of such endomorphisms are given.

§2. Endomorphisms on $\mathscr{B}(X, K)$

In the sequel, we shall consider only commutative semirings K (see Definition 2.1 in [1]) with operations $\oplus$ and $\odot$ having neutral elements $\mathbf{0}$ and $\mathbf{1}$ respectively and satisfying the following conditions:

CONDITION 2.1. *The operation $\oplus$ is assumed to be idempotent, i.e., we have $a \oplus a = a$ for all $a \in K$.*

1991 *Mathematics Subject Classification.* Primary 16Y60, 47B38.

CONDITION 2.2. *The set* $K \setminus \{\mathbf{0}\}$ *is a group under the multiplication* $\odot$.

CONDITION 2.3 (stabilization). *For arbitrary* $\lambda, b \in K$ *such that* $\lambda \leqslant \mathbf{1}$, *and* $b \neq \mathbf{0}$ *the sequence* $\{\lambda^m \oplus b\}_{m=1}^{\infty}$ *stabilizes.*

CONDITION 2.4 (algebraic completeness). *The semiring* K *is algebraically closed, i.e., the equation* $x^m = x \odot x \odot \cdots \odot x = a$ *has a solution* $x \in K$ *for an arbitrary* $a \in K$ *and an arbitrary positive integer* m.

CONDITION 2.5. *The semiring* K *as a partially ordered set (the idempotent operation* $\oplus$ *permits us to endow* K *with a partial order) is a complete lattice, i.e., every bounded set in* K *has a supremum and an infimum.*

The idempotent operation $\oplus$ induces a *partial order* $\leqslant$ in the set K by the rule

$$a \leqslant b \iff a \oplus b = b.$$

This partial order in K possesses the following properties:

(1) The element $\mathbf{0} \in K$ is the smallest one, therefore all elements from K are “nonnegative”;
(2) $a \leqslant b \Rightarrow a \oplus c \leqslant b \oplus c$, $\forall c \in K$;
(3) $a \leqslant b \Rightarrow a \odot c \leqslant b \odot c$, $\forall c \in K$;
(4) $a \oplus b = \sup(a, b)$, $\forall a, b \in K$.

ASSERTION 2.1 [1]. *If a semiring* K *satisfies Conditions* 2.1, 2.2 *and* 2.4, *then the equation* $x^m = a$ *has a unique solution for an arbitrary* $a \in K$ *and an arbitrary positive integer* m.

EXAMPLE 2.1. The semiring $\mathscr{R} = \mathbb{R}^1 \cup \{-\infty\}$ with the operations $\oplus =$ max and $\odot = +$ (see [1]).

EXAMPLE 2.2. Let $K = \mathbb{R}^1 \cup \{+\infty\}$, $a \oplus b = \min(a, b)$, $a \odot b = a + b$, $\mathbf{0} = +\infty$, $\mathbf{1} = 0$. In this semiring the partial order induced by the operation $\oplus$ is linear and is opposite to the usual order in $\mathbb{R}^1$.

In Definition 3.1 from [1], the concept of a *semimodule* M *over a semiring* K was introduced. Further we shall use this definition.

Everywhere below we shall assume that the “addition” operation $\oplus$ in M is idempotent, i.e., for all elements $m \in M$ the identity $m \oplus m = m$ is true.

As in the case of semirings, the idempotent operation $\oplus$ induces a *partial order* $\leqslant$ in the set M according to the rule:

$$m_1 \leqslant m_2 \iff m_1 \oplus m_2 = m_2.$$

EXAMPLE 2.3. Let X be an arbitrary set. Denote by $\mathscr{B}(X, K)$ the set of bounded mappings from X into a semiring K (i.e., the mapping $f\colon X \to K$ belongs to the set $\mathscr{B}(X, K)$ iff there exists an element $M_f \in K$ such that $f(x) \leqslant M_f$ for all $x \in X$). The set $\mathscr{B}(X, K)$ is a semimodule over the semiring K with respect to the operations

$$(f \oplus g)(x) = f(x) \oplus g(x), \qquad f, g \in \mathscr{B}(X, K),\ x \in X,$$
$$(\lambda \odot f)(x) = \lambda \odot f(x), \qquad f \in \mathscr{B}(X, K),\ x \in X,\ \lambda \in K.$$

The partial order in K (induced by the idempotent addition $\oplus$) induces a partial order in $\mathscr{B}(X, K)$ in the following way:

$$f \leqslant g \iff f(x) \leqslant g(x), \ \forall x \in X.$$

It is not difficult to verify that the set $\mathscr{B}(X, K)$ is a complete lattice with respect to this partial order.

If the set X is finite and consists of n elements, then the corresponding semimodule $\mathscr{B}(X, K)$ is a collection of "vectors" $(a_1, a_2, \dots, a_n)$ $(a_i \in K)$ and is denoted by K^n.

ASSERTION 2.2 [1]. *Let X be an arbitrary set, K a commutative semiring with an idempotent operation $\oplus$, and $\mathscr{B}(X, K) = V_1 \dotplus V_2$, where $\dotplus$ is the direct sum, V_i $(i = 1, 2)$ is a subsemimodule of $\mathscr{B}(X, K)$ (i.e., every element f can be uniquely expressed in the form $f = v_1 \oplus v_2$, where $v_i \in V_i$ $(i = 1, 2)$). Then there exists a decomposition $X = X_1 \cup X_2$ $(X_1 \cap X_2 = \varnothing)$ such that $V_i = \mathscr{B}(X_i, K)$, and $v_i \in \mathscr{B}(X_i, K)$ is extended to X by zero outside X_i.*

EXAMPLE 2.4. Let X be a topological space and $K = \mathscr{R}$ (see Example 2.1). Then the set $C(X, \mathscr{R})$ of continuous functions on X with values in $\mathscr{R}$ is a semimodule over the semiring $\mathscr{R}$. If X is a compact set, then the semimodule $C(X, \mathscr{R})$ is a subsemimodule of the semimodule $\mathscr{B}(X, K)$ (in the obvious sense).

EXAMPLE 2.5. Let $X = \mathbb{R}^n$ and $K = \mathscr{R}$. Then the set $\mathrm{Conc}(X, \mathscr{R})$ of concave functions on X with values in $\mathscr{R}$ is a semimodule over the semiring $\mathscr{R}$.

DEFINITION 2.1. A set of elements $\{m_u \mid u \in U, m_u \in M\}$ is called a *system of generators* in M, if every element $m \in M$ can be written as follows:

$$m = \bigoplus_{u \in U} \alpha(u) \odot m_u = \sup_{u \in U} \alpha(u) \odot m_u,$$

where $\alpha(u) \in K$ for all $u \in U$.

EXAMPLE 2.6. The set of function $\{\delta_u(x)\}_{u \in U}$ of the form

$$\delta_u(x) = \begin{cases} \mathbf{1} & \text{for } x = u, \\ \mathbf{0} & \text{for } x \neq u \end{cases}$$

is a system of generators in $\mathscr{B}(X, K)$.

EXAMPLE 2.7. Let $X = \mathbb{R}^n$, $K = \mathscr{R}$. Then the set of linear functions $\{(\beta, x)\}_{\beta \in \mathbb{R}^n}$ is a system of generators in the semimodule $\mathrm{Conc}(X, \mathscr{R})$, since every element $f \in \mathrm{Conc}(X, \mathscr{R})$ has the following form

$$f(x) = \sup_\beta \alpha(\beta) \odot (\beta, x) = \sup_\beta \{\alpha(\beta) + (\beta, x)\},$$

where we can take the function $\gamma(\beta) = \inf_x \{f(x) - (\beta, x)\}$ to be the function $\alpha(\beta)$ (this follows from the properties of the Legendre transform).

Example 2.7 has a generalization: suppose a semimodule M over a semiring K satisfies the following

Condition 2.6 (absence of infinite elements). *Let $m, a \in M$ and $\lambda_\alpha \in K$ (α belongs to some set of indices I) be such that:*

i) *the set $\{\lambda_\alpha\}_{\alpha\in I}$ is an unbounded set in K;*
ii) *$m \geqslant \lambda_\alpha \odot a$ for all $\alpha \in I$.*

Then $a = \mathbf{0}$.

Proposition 2.1. *Let $\{m_u\}_{u\in U}$ be a system of generators of the semimodule M (over the semiring K). For an arbitrary element $m \in M$ we determine a function $\alpha(u)$ on the set U with values in K in the following way:*

$$\alpha(u) = \sup\{\lambda : m \geqslant \lambda \odot m_u\}.$$

Then the equality

$$m = \bigoplus_{u\in U} \alpha(u) \odot m_u = \sup_{u\in U} \alpha(u) \odot m_u$$

holds.

Proof. Since the set $V_u = \{\lambda : m \geqslant \lambda \odot m_u\}$ is nonempty (it contains $\mathbf{0}$) and is bounded (this follows from Condition 2.6), the function $\alpha(u)$ is well defined. The inequality $m \geqslant \lambda \odot m_u$, which holds for all $\lambda \in V_u$, and the axioms of semimodules yield the inequality

$$m \geqslant \left(\sup_{\lambda\in V_u} \lambda\right) \odot m_u = \alpha(u) \odot m_u.$$

Thus the element m is an upper bound of the set $\{\alpha(u) \odot m_u\}_{u\in U}$. We shall show that this upper bound is a supremum. Let m_1 be an element from M such that $m_1 \geqslant \alpha(u) \odot m_u$ for all $u \in U$. It is necessary to prove that $m_1 \geqslant m$. Since $\{m_u\}_{u\in U}$ is a system of generators in M and $m \in M$, we see that there exist elements $\bar{\alpha}(u) \in M$ for which we have

$$m = \sup_{u\in U} \bar{\alpha}(u) \odot m_u.$$

From this equality it follows that for all $u \in U$ the inequality $m \geqslant \bar{\alpha}(u) \odot m_u$ is true. And so $\bar{\alpha}(u) \in V_u$. Therefore $\bar{\alpha}(u) \leqslant \alpha(u)$. Then

$$m_1 \geqslant \alpha(u) \odot m_u \geqslant \bar{\alpha}(u) \odot m_u.$$

Passing to the supremum in the last inequality, we obtain

$$m_1 \geqslant \sup_{u\in U} \bar{\alpha}(u) \odot m_u = m.$$

The proof is over.

Corollary 2.1. *Assume that $\{m_u\}_{u\in U}$ is a system of generators of the semimodule M over the semiring K and m is an arbitrary element from M. Define a function $\alpha(u)$ on the set U with values in K as follows:*

$$\alpha(u) = \sup\{\lambda : m \geqslant \lambda \odot m_u\}.$$

Then we have the relation

$$\alpha(u) = \sup \left\{ \beta \in \mathscr{B}(U, K) : m = \sup_{u \in U} \beta(u) \odot m_u \right\}.$$

§3. Endomorphisms

If M_1 and M_2 are semimodules over a semiring K, then the mapping $\Lambda: M_1 \to M_2$ is a *homomorphism of semimodules* whenever the equality

$$\Lambda(\lambda \odot f \oplus \mu \odot g) = \lambda \odot \Lambda f \oplus \mu \odot \Lambda g$$

holds for all $f, g \in M_1$ and $\lambda, \mu \in K$.

Homomorphisms from M into M are called *endomorphisms*. We shall denote the set of all endomorphisms acting in M by $\operatorname{End}(M)$.

For simplicity everywhere below we shall assume that $M = \mathscr{B}(X, K)$ and $\{\delta_y(x) \mid y \in X\}$ is a system of generators of $\mathscr{B}(X, K)$ (see Examples 2.3, 2.6).

DEFINITION 3.1. An endomorphism $\Lambda: \mathscr{B}(X, K) \to \mathscr{B}(X, K)$ is called *monotonically continuous* if it satisfies the following condition:

$$\Lambda\left\{ \sup_{\alpha \in I} f_\alpha \right\} = \sup_{\alpha \in I} \Lambda f_\alpha, \tag{1}$$

where I is an arbitrary parameter set, $f_\alpha \in \mathscr{B}(X, K)$ for all $\alpha \in I$, and the set $\{f_\alpha \mid \alpha \in I\}$ is bounded in $\mathscr{B}(X, K)$.

Let $K(x, y) = \Lambda \delta_y(x)$. Then for an arbitrary $f \in \mathscr{B}(X, K)$ from equality (1) we have

$$(\Lambda f)(x) = \sup_{y \in X} f(y) \odot K(x, y). \tag{2}$$

The function $K(x, y)$ will be called the *kernel* of Λ.

ASSERTION 3.1. *Suppose $\Lambda: \mathscr{B}(X, K) \to \mathscr{B}(X, K)$ is an endomorphism satisfying relation* (1). *Then its kernel $K(x, y)$ is a bounded function on $X \times X$.*

PROOF. Let $f_0(x) \equiv \mathbf{1}$. Then for all $x \in X$ we have $(\Lambda f_0)(x) \leqslant M(f_0)$, where $M(f_0)$ is some constant. But $f_0(x) = \sup_{y \in X} \delta_y(x)$. And so, using (2), we obtain the identity

$$\Lambda f_0(x) = \sup_{y \in X} \Lambda \delta_y(x) = \sup_{y \in X} K(x, y).$$

Therefore the inequality $\sup_{x \in X} \sup_{y \in X} K(x, y) \leqslant M(f_0)$ holds for all $x \in X$.

DEFINITION 3.2. Let $\Lambda: \mathscr{B}(X, K) \to \mathscr{B}(X, K)$ be a monotonically continuous endomorphism and $K(x, y)$ be its kernel. Suppose

$$\operatorname{Tr}(\Lambda) = \sup_{x \in X} K(x, x), \qquad \|\Lambda\| = \sup_{x, y \in X} K(x, y)$$

and define the terms of the sequences $\{\underline{\lambda}_k\}_{k=1}^{\infty}$ and $\{\bar{\lambda}_k\}_{k=1}^{\infty}$ of elements from K as the roots of the equations $\underline{\lambda}_k^k = \operatorname{Tr}(\Lambda^k)$ and $\bar{\lambda}_k^k = \|\Lambda^k\|$ respectively.

PROPOSITION 3.1. *Let* $\Lambda\colon \mathscr{B}(X, K) \to \mathscr{B}(X, K)$ *be a monotonically continuous endomorphism. Then for all* $f \in \mathscr{B}(X, K)$ *we have the inequality*

$$\|\Lambda f\| \leqslant \|\Lambda\| \odot \|f\|,$$

where $\|f\| = \sup_{x\in X} f(x)$.

PROPOSITION 3.2. *For all natural* p *and* q, *the sequences* $\{\underline{\lambda}_k\}_{k=1}^{\infty}$ *and* $\{\bar{\lambda}_k\}_{k=1}^{\infty}$ *satisfy the following relations*:

$$\text{i) } \underline{\lambda}_p \leqslant \underline{\lambda}_{pq}, \qquad \text{ii) } \bar{\lambda}_{pq} \leqslant \bar{\lambda}_p, \qquad \text{iii) } \underline{\lambda}_p \leqslant \bar{\lambda}_q.$$

PROOF. i) Denote by $K^n(x, y)$ the kernel of the endomorphism Λ^n. We have the relations

$$\begin{aligned}
\underline{\lambda}_{pq}^{pq} &= \operatorname{Tr}(\Lambda^{pq}) = \operatorname{Tr}((\Lambda^p)^q) \\
&= \sup_{x\in X}\ \sup_{y_1, y_2, \dots, y_{q-1}} K^p(x, y_1) \odot K^p(y_1, y_2) \odot \cdots \odot K^p(y_{q-1}, x) \\
&\geqslant \sup_{x\in X} K^p(x, x) \odot K^p(x, x) \odot \cdots \odot K^p(x, x) \\
&= \sup_{x\in X}(K^p(x, x))^q = \Big(\sup_{x\in X} K^p(x, x)\Big)^q = (\operatorname{Tr}(\Lambda^p))^q = \underline{\lambda}_p^{pq},
\end{aligned}$$

whence inequality i) follows.

ii) We have the inequalities

$$\begin{aligned}
\bar{\lambda}_{p+q}^{p+q} &= \|\Lambda^{p+q}\| = \sup_{x, y\in X}\ \sup_{y_1} K^p(x, y_1) \odot K^p(y_1, y) \\
&\leqslant \sup_{x, y\in X} K^p(x, y) \odot \sup_{x, y\in X} K^q(x, y) = \|\Lambda^p\| \odot \|\Lambda^q\| = \bar{\lambda}_p^p \odot \bar{\lambda}_p^q.
\end{aligned}$$

The relation $\bar{\lambda}_{p+q}^{p+q} \leqslant \bar{\lambda}_p^p \odot \bar{\lambda}_q^q$ yields the inequality

$$\bar{\lambda}_{pq}^{pq} = \bar{\lambda}_{p+p+\cdots+p}^{p+p+\cdots+p} \leqslant \bar{\lambda}_p^p \odot \bar{\lambda}_p^p \odot \cdots \odot \bar{\lambda}_p^p = \bar{\lambda}_p^{pq},$$

whence inequality ii) follows.

Inequalities i) and ii) immediately imply inequality iii).

DEFINITION 3.3. The elements

$$\lambda_{\min}(\Lambda) = \sup_{k=1, 2, \dots} \{\underline{\lambda}_k\}, \qquad \lambda_{\max}(\Lambda) = \inf_{k=1, 2, \dots} \{\bar{\lambda}_k\}$$

are called *spectral elements* of the endomorphism Λ.

PROPOSITION 3.3. *The spectral elements* $\lambda_{\min}(\Lambda)$, $\lambda_{\max}(\Lambda)$ *of an arbitrary monotonically continuous endomorphism* $\Lambda\colon \mathscr{B}(X, K) \to \mathscr{B}(X, K)$ *satisfy the inequality*

$$\lambda_{\min}(\Lambda) \leqslant \lambda_{\max}(\Lambda).$$

The proof immediately follows from assertion iii) of Proposition 3.2.

COROLLARY 3.1. *Let* $\Lambda\colon \mathscr{B}(X, K) \to \mathscr{B}(X, K)$ *be a monotonically continuous endomorphism such that* $\underline{\lambda}_q = \bar{\lambda}_p$ *for some* $p, q \in \mathbb{N}$. *Then* $\lambda_{\min}(\Lambda) = \lambda_{\max}(\Lambda)$.

EXAMPLE 3.1. Let $\Lambda\colon \mathscr{B}(X, K) \to \mathscr{B}(X, K)$ be a monotonically continuous endomorphism. Assume that its kernel $K(x, y)$ is symmetric, i.e., $K(x, y) = K(y, x)$ for all $x, y \in X$. Then $\underline{\lambda}_2^2 = \bar{\lambda}_1^2$, and therefore, $\lambda_{\min}(\Lambda) = \lambda_{\max}(\Lambda)$.

Suppose X, Y are arbitrary sets and $Y \subseteq X$. Then there exists an imbedding $T\colon \mathscr{B}(Y, K) \to \mathscr{B}(X, K)$ that assigns to every function $f \in \mathscr{B}(Y, K)$ its extension to X by zero outside Y. In the sequel, we shall identify $\mathscr{B}(Y, K)$ with its image $T\mathscr{B}(Y, K) \subseteq \mathscr{B}(X, K)$.

PROPOSITION 3.4. *Let* $\Lambda\colon \mathscr{B}(X, K) \to \mathscr{B}(X, K)$ *be a monotonically continuous endomorphism and* $f \in \mathscr{B}(X, K)$ *be its eigenelement* $(f \not\equiv \mathbf{0})$. *Suppose* $Y = \operatorname{supp} f$. *Then the semimodule* $\mathscr{B}(Y, K)$ *is invariant with respect to the endomorphism* Λ, *i.e., for all* $g \in \mathscr{B}(Y, K)$ *we have* $\Lambda g \in \mathscr{B}(Y, K)$.

PROOF. Denote by λ_0 an element such that $\Lambda f = \lambda_0 \odot f$ and assume that $Y^{(1)} = X \setminus Y$. Then $\mathscr{B}(X, K) = \mathscr{B}(Y, K) \dot{+} \mathscr{B}(Y^{(1)}, K)$, where $\dot{+}$ is the direct sum, and the endomorphism Λ can be written in matrix form as follows:

$$\Lambda = \begin{pmatrix} \Lambda_{11} & \Lambda_{12} \\ \Lambda_{21} & \Lambda_{22} \end{pmatrix},$$

where the endomorphisms Λ_{ij} act in the corresponding subspaces. In such a decomposition the element f has the following form:

$$f = (f_1, \mathbf{0}), \quad \text{where } f_1(x) > \mathbf{0} \text{ for all } x \in Y.$$

The condition $\Lambda f = \lambda_0 \odot f$ yields the identity $(\Lambda_{21} f_1)(x) \equiv \mathbf{0}$ for all $x \in Y$ and this last identity implies $\Lambda_{21} = \mathbf{0}$. This equality is equivalent to the invariance of the semimodule $\mathscr{B}(Y, K)$.

PROPOSITION 3.5. *Let* $\Lambda\colon \mathscr{B}(X, K) \to \mathscr{B}(X, K)$ *be a monotonically continuous endomorphism and* $f \in \mathscr{B}(X, K)$ *its eigenelement with eigenvalue* λ_0 $(f \not\equiv \mathbf{0})$. *Suppose* $Y = \operatorname{supp} f$, Λ_f *is the restriction of the endomorphism* Λ *to the semimodule* $\mathscr{B}(Y, K)$, *and* $\lambda_{\min}(\Lambda_f)$, $\lambda_{\max}(\Lambda_f)$ *are the corresponding spectral elements. Then we have the inequalities*

$$\lambda_{\min}(\Lambda_f) \leqslant \lambda_0 \leqslant \lambda_{\max}(\Lambda_f).$$

PROOF. To prove the inequality $\lambda_{\min}(\Lambda_f) \leqslant \lambda_0$ it is obviously sufficient to prove the inequalities $\underline{\lambda}_p \leqslant \lambda_0$ for all $p = 1, 2, \dots$, where $\underline{\lambda}_p^p = \operatorname{Tr}(\Lambda_f^p)$. It is not difficult to verify that the kernel $K_f^p(x, y)$ of the endomorphism Λ_f^p can be expressed by means of the kernel $K_f(x, y)$ of the endomorphism Λ_f as follows:

$$K_f^p(x, y) = \sup_{x_1, \dots, x_{p-1} \in Y} [K_f(x, x_1) \odot K_f(x_1, x_2) \odot \dots \odot K_f(x_{p-1}, y)].$$

And so to prove the inequality $\lambda_p \leqslant \lambda_0$ it is appropriate to prove the inequality

$$K_f(x, x_1) \odot K_f(x_1, x_2) \odot \cdots \odot K_f(x_{p-1}, x) \leqslant \lambda_0^p \tag{3}$$

for all points $x_1, x_2, \ldots, x_{p-1} \in Y$.

For all $z, y \in Y$ the inequality

$$K_f(z, y) \odot f(y) \leqslant \lambda_0 \odot f(z)$$

is true. Supposing $z = x_i$, $y = x_{i+1}$ $(i = 0, 1, \ldots, p-1)$, $x_0 = x_p = x$, from the last inequality we have

$$K_f(x_i, x_{i+1}) \odot f(x_{i+1}) \leqslant \lambda_0 \odot f(x_i), \qquad i = 0, 1, \ldots, p-1.$$

Multiplying these inequalities, we obtain

$$K_f(x, x_1) \odot \cdots \odot K_f(X_{p-1}, x) \odot f(x) \odot \cdots \odot f(x_{p-1})$$
$$\leqslant \lambda_0^p \odot f(x) \odot \cdots \odot f(x_{p-1}).$$

Hence the condition $f(y) \neq \mathbf{0}$ for all $y \in Y$ and the cancellation condition (Condition 2.1 from [1]) imply inequality (3). Thus the inequality $\lambda_{\min}(\Lambda_f) \leqslant \lambda_0$ is proved.

And now we prove the inequality $\lambda_0 \leqslant \lambda_{\max}(\Lambda_f)$. To do this, it is sufficient to prove the inequalities $\lambda_0 \leqslant \bar{\lambda}_p$ for all $p = 1, 2, \ldots$. For every element $g \in \mathscr{B}(Y, K)$, we have the estimate

$$\|\Lambda_f^p g\| = \sup_{x, y \in Y} (K_f^p(x, y) \odot g(y))$$
$$\leqslant \sup_{x, y \in Y} K_f^p(x, y) \odot \sup_{x \in Y} g(y) = \|\Lambda_f^p\| \, \|g\| = \bar{\lambda}_p^p \odot \|g\|.$$

Substituting $g = f$ in this inequality and taking into account the fact that $\Lambda_f^p = \lambda_0^p \odot f$, we obtain the inequality

$$\lambda_0^p \odot \|f\| = \|\Lambda_f^p\| \leqslant \bar{\lambda}_p^p \odot \|f\|.$$

Therefore, from the condition $f(y) \neq \mathbf{0}$ for all $y \in Y$ and the cancellation condition the required inequality follows. Thus the proof is complete.

COROLLARY 3.2. *Let $\lambda_0 \in K$ be an arbitrary eigenelement of an endomorphism*

$$\Lambda \colon \mathscr{B}(X, K) \to \mathscr{B}(X, K).$$

Then $\lambda_0 \leqslant \lambda_{\max}(\Lambda)$.

PROPOSITION 3.6. *Let $\Lambda \colon \mathscr{B}(X, K) \to \mathscr{B}(X, K)$ be a monotonically continuous endomorphism and $\{\bar{\lambda}_k\}_{k=1}^{\infty}$ be the sequence determined by Definition 3.2. Then for an arbitrary positive integer p there exists an element $M_p \in K$ such that for all indices $k = 0, 1, \ldots$ and functions $f \in \mathscr{B}(X, K)$ we have the inequality*

$$\|\Lambda^k f\| \leqslant M_p \odot \bar{\lambda}_p^k \odot \|f\|. \tag{4}$$

PROOF. From Proposition 3.1 it follows that for every fixed index $p \in \mathbb{N}$ one can choose an element $M_p \in K$ such that $\forall k = 0, 1, 2, \ldots, p-1$ and functions $f \in \mathscr{B}(X, K)$, the inequality

$$\|\Lambda^k f\| \leqslant M_p \odot \bar{\lambda}_p^k \odot \|f\|$$

takes place.

Let now $k = mp + r$ $(0 \leqslant r \leqslant p-1)$ be an arbitrary index. Then using Proposition 3.1, we obtain the series of inequalities

$$\begin{aligned}\|\Lambda^k f\| &= \|\Lambda^{mp+r} f\| = \|\Lambda^{mp} \circ \Lambda^r f\| \leqslant \|\Lambda^{mp}\| \odot \|\Lambda^r f\| \\ &\leqslant \|\Lambda^p\|^m \odot \|\Lambda^r f\| \leqslant M_p \odot (\bar{\lambda}_p^p)^m \odot \bar{\lambda}_p^r \odot \|f\| = M_p \odot \bar{\lambda}_p^k \odot \|f\|,\end{aligned}$$

whence inequality (4) follows. Proposition 3.6 is proved.

COROLLARY 3.3. *Let $\Lambda\colon \mathscr{B}(X, K) \to \mathscr{B}(X, K)$ be a monotonically continuous endomorphism such that $\lambda_{\max}(\Lambda) = \bar{\lambda}_p$ for some $p \in \mathbb{N}$. Then there exists an element $M \in K$ such that for all indices $k = 0, 1, \ldots$ and functions $f \in \mathscr{B}(X, K)$ the inequality*

$$\|\Lambda^k f\| \leqslant M \odot \lambda_{\max}^k(\Lambda) \odot \|f\|$$

holds.

PROPOSITION 3.7. *Let $\Lambda\colon \mathscr{B}(X, K) \to \mathscr{B}(X, K)$ be a monotonically continuous endomorphism and $\lambda_{\max}(\Lambda) \in K$ be the corresponding spectral element. Denote by U_Λ the set of elements $\lambda \in K$ for which there exists an element $M = M(\lambda) \in K$ such that for all indices $k = 0, 1, \ldots$ and functions $f \in \mathscr{B}(X, K)$ the inequality*

$$\|\Lambda^k f\| \leqslant M \odot \lambda^k \odot \|f\|$$

holds. Then $\lambda_{\max}(\Lambda) = \inf\{\lambda \mid \lambda \in U_\Lambda\}$.

PROOF. Inequality (4) implies the inclusions $\bar{\lambda}_p \in U_\Lambda$ for all $p \in \mathbb{N}$. Since $\lambda_{\max}(\Lambda) = \inf_p \bar{\lambda}_p$, we have $\lambda_{\max}(\Lambda) \geqslant \inf\{\lambda \mid \lambda \in U_\Lambda\}$. Now we prove the opposite inequality $\lambda_{\max}(\Lambda) \leqslant \inf\{\lambda \mid \lambda \in U_\Lambda\}$. Suppose $\lambda_0 \in U_\Lambda$. Then there exists an element $M \in K$ such that for all $p = 1, 2, \ldots$ and $f \in \mathscr{B}(X, K)$ the inequality

$$\|\Lambda^p f\| \leqslant M \odot \lambda_0^p \odot \|f\| \tag{5}$$

holds. We have $K(x, y) = \Lambda\delta_y(x)$, where

$$\delta_y(x) = \begin{cases} \mathbf{1} & \text{for } x = y, \\ \mathbf{0} & \text{otherwise.} \end{cases}$$

And so

$$\|\Lambda\| = \sup_{x, y \in X} K(x, y) = \sup_{y \in X} \sup_{x \in X} K(x, y) = \sup_{y \in X} \sup_{x \in X} \Lambda\delta_y(x) = \sup_{y \in X} \|\Lambda\delta_y\|.$$

Hence

$$\lambda_p^p = \|\Lambda^p\| = \sup_{y \in X} \|\Lambda^p \delta_y\|.$$

From inequality (5) it follows that

$$\lambda_p^p = \sup_{y \in X} \|\Lambda^p \delta_y\| \leqslant M \odot \lambda_0^p \odot \|\delta_y\|.$$

Thus for all $p \in \mathbb{N}$ the inequality $\lambda_p^p \leqslant M \odot \lambda_0^p$ holds. Taking into account the inequality $\lambda_{\max}(\Lambda) \leqslant \lambda_p$ for all $p \in \mathbb{N}$, we obtain the following relation: $\lambda_m^p(\Lambda) \leqslant M \odot \lambda_0^p$. The stabilization condition and the last inequality yield $\lambda_{\max}(\Lambda) \leqslant \lambda_0$, whence the required inequality follows. The proof is complete.

DEFINITION 3.4. Let $\Lambda: \mathscr{B}(X, K) \to \mathscr{B}(X, K)$ be a monotonically continuous endomorphism and $K(x, y)$ its kernel. An endomorphism Λ^* acting in $\mathscr{B}(X, K)$ is called *adjoint* to the endomorphism Λ if

$$(\Lambda^* F)(x) = \sup_{y \in X} K(y, x) \odot f(y).$$

PROPOSITION 3.8. *Let $\Lambda: \mathscr{B}(X, K) \to \mathscr{B}(X, K)$ be a monotonically continuous endomorphism and Λ^* be adjoint to Λ. Assume that elements λ, $\lambda^* \in K$ $(\lambda \neq \lambda^*)$ and f, $f^* \in \mathscr{B}(X, K)$ are such that $\Lambda f = \lambda \odot f$, $\Lambda^* f^* = \lambda^* \odot f^*$ and $\|f\| = \|f^*\| = 1$. Then $\operatorname{supp} f \cap \operatorname{supp} f^* = \varnothing$.*

PROOF. It is sufficient to prove that $\sup_{y \in X} f(y) \odot f^*(y) = \mathbf{0}$. We have

$$\begin{aligned}\lambda \odot \sup_{y \in X} f(y) \odot f^*(y) &= \sup_{y \in X} \Lambda f(y) \odot f^*(y) \\ &= \sup_{y \in X} f(y) \odot \Lambda^* f^*(y) = \lambda^* \odot \sup_{y \in X} f(y) \odot f^*(y).\end{aligned}$$

From the condition $\lambda \neq \lambda^*$, the cancellation condition, and the equality

$$\lambda \odot \sup_{y \in X} f(y) \odot f^*(y) = \lambda^* \odot \sup_{y \in X} f(y) \odot f^*(y),$$

obtained previously, the required equality $\sup_{y \in X} f(y) \odot f^*(y) = \mathbf{0}$ now follows. Therefore Proposition 3.8 is proved.

§4. Indecomposable endomorphisms acting in $\mathscr{B}(X, K)$

DEFINITION 4.1. Let $\Lambda: \mathscr{B}(X, K) \to \mathscr{B}(X, K)$ be a monotonically continuous endomorphism and $K(x, y)$ be its kernel. The endomorphism Λ is called *decomposable*, if there exists a decomposition $X = X_1 \cup X_2$ $(X_i \neq \varnothing$, $X_1 \cap X_2 = \varnothing)$ such that $K(x, y) = \mathbf{0}$ for all $x \in X_1$, $y \in X_2$, and *indecomposable* otherwise.

REMARK 4.1. If X is a finite set, then Definition 4.1 is equivalent to the one from [1].

PROPOSITION 4.1. *Let $\Lambda: \mathscr{B}(X, K) \to \mathscr{B}(X, K)$ be a monotonically continuous endomorphism and $K(x, y)$ its kernel. Λ is indecomposable if and* *only if for any point* $x, y \in X$ $(x \neq y)$ *one of the following conditions is satisfied*:

i) $K(x, y) \neq \mathbf{0}$;

ii) *there exists a sequence of points* $x_1, x_2, \dots, x_s \in X$ *such that*

$$K(x, x_1) \odot K(x_1, x_2) \odot \cdots \odot K(x_s, y) \neq \mathbf{0}.$$

PROOF. The sufficiency of the conditions is obvious. Let us prove its necessity. Assume that there exist points $x_0, y_0 \in X$ $(x_0 \neq y_0)$ such that $K(x_0, y_0) = \mathbf{0}$ and $K(x_0, x_1) \odot K(x_1, x_2) \odot \cdots \odot K(x_s, y_0) = \mathbf{0}$ for any sequence of points $x_1, x_2, \dots, x_s \in X$. We show that in such a case Λ is decomposable. Define the sets X_1 and X_2 in the following way:

$$X_1 = \{x_0\} \cup \{x \in X \mid \exists x_1, x_2, \dots, x_s \in X : \\ K(x_0, x_1) \odot K(x_1, x_2) \odot \cdots \odot K(x_s, x) \neq \mathbf{0}\},$$

$$X_2 = X \setminus X_1.$$

Then every set is nonempty (since $x_0 \in X_1$, $y_0 \in X_2$), $X_1 \cap X_2 = \varnothing$ and $X = X_1 \cup X_2$. Let us show that $K(x, y) = \mathbf{0}$ holds for all $x \in X_1$, $y \in X_2$. Indeed, let $K(x, y) \neq \mathbf{0}$ for some $x \in X_1$, $y \in X_2$. Since $x \in X_1$, we see that there exists a sequence of points $x_1, x_2, \dots, x_s \in X$ such that $K(x_0, x_1) \odot K(x_1, x_2) \odot \cdots \odot K(x_s, x) \neq \mathbf{0}$. Then

$$K(x_0, x_1) \odot K(x_1, x_2) \odot \cdots \odot K(x_s, x) \odot K(x, y) \neq \mathbf{0},$$

and, therefore, $y \in X_1$, whence the inclusion $y \in X_1 \cap X_2$ follows. But this inclusion contradicts the choice of X_1 and X_2. Thus $K(x, y) = \mathbf{0}$ for all $x \in X_1$, $y \in X_2$ and so Λ is decomposable. The proof is complete.

PROPOSITION 4.2. *Let* $\Lambda\colon \mathscr{B}(X, K) \to \mathscr{B}(X, K)$ *be a monotonically continuous endomorphism. Assume that its kernel* $K(x, y)$ *is symmetric, i.e.,* $K(x, y) = K(y, x)$ *for all* $x, y \in X$. *Then* Λ *has at most one eigenelement* λ_Λ *which coincides with* $\lambda_{\min}$ ($\lambda_{\min}$ *is the spectral element of* Λ), *if it exists.*

The proof immediately follows from Propositions 3.4, 3.5 and Example 3.1.

PROPOSITION 4.3. *Let* $\Lambda\colon \mathscr{B}(X, K) \to \mathscr{B}(X, K)$ *be a monotonically continuous endomorphism and* $f(x) \in \mathscr{B}(X, K)$ $(\|f\| = \mathbf{1})$ *be an element such that* $\Lambda f \leqslant \lambda \odot f$ *for some* $\lambda \in K$. *Then* $f(x) \neq \mathbf{0}$ *for all* $x \in X$.

PROOF. Assume the contrary. Let the set $X_1 = \{x \in X \mid f(x) = \mathbf{0}\}$ be nonempty. Since $\|f\| = 1$, the set $X_2 = X \setminus X_1$ is also nonempty and for all $x \in X_1$ the following equality is satisfied:

$$(\Lambda f)(x) = \sup_{y \in X} K(x, y) \odot f(y) = \mathbf{0}.$$

And so $\sup_{y \in X_2} K(x, y) \odot f(y) = \mathbf{0}$, from where for all $x \in X_1$ and $y \in X_2$ we have $K(x, y) \odot f(y) = \mathbf{0}$. But $f(y) \neq \mathbf{0}$ for $y \in X_2$. Hence the cancellation condition yields $K(x, y) = \mathbf{0}$ for all $x \in X_1$ and $y \in X_2$ and this contradicts the indecomposability property of Λ.

PROPOSITION 4.4. *Let* $\Lambda: \mathscr{B}(X, K) \to \mathscr{B}(X, K)$ *be a monotonically continuous endomorphism and* Λ^* *be its adjoint endomorphism. Assume that elements* $\lambda, \lambda^* \in K$ *and* $f, f^* \in \mathscr{B}(X, K)$ *are such that* $\Lambda f = \lambda \odot f$, $\Lambda^* f^* = \lambda^* \odot f^*$ *and* $\|f\| = \|f^*\| = 1$. *Then* $\lambda = \lambda^*$.

The proof immediately follows from Propositions 3.8 and 4.3.

COROLLARY 4.1. *Let* $\Lambda: \mathscr{B}(X, K) \to \mathscr{B}(X, K)$ *be a monotonically continuous indecomposable endomorphism. Assume that* Λ *has two eigenvalues* $\lambda, \lambda_1 \in K$ *that differ. Then its adjoint endomorphism* Λ^* *does not have any eigenvalues.*

§5. Uniformly indecomposable endomorphisms acting in $\mathscr{B}(X, K)$

DEFINITION 5.1. Every monotonically continuous endomorphism Λ acting in $\mathscr{B}(X, K)$ is called *uniformly indecomposable*, if its kernel satisfies the following condition

$$\inf_{x, y} \sup_{k=0,1,\dots} (\lambda_{\min}^{k-1} \odot K^k(x, y)) = c > \mathbf{0}, \tag{6}$$

where $K^k(x, y)$ is the kernel of the endomorphism Λ^k and $\lambda_{\min} = \lambda_{\min}(\Lambda)$ is a spectral element of Λ.

REMARK 5.1. Condition (6) is satisfied if, for instance, for all $x, y \in X$ we have the inequality $K(x, y) > c_0 \neq \mathbf{0}$. If X is a finite set, then uniform indecomposability is equivalent to indecomposability.

REMARK 5.2. An arbitrary uniformly indecomposable endomorphism is obviously indecomposable. It follows from the next example (due to A. A. Tarashchan) that the converse assertion, generally speaking, is not true.

EXAMPLE 5.1. Let $K = \mathscr{R}$ (see Example 2.1), $X = \mathbb{N}$. The endomorphism Λ is defined by an infinite matrix of the form

$$\begin{pmatrix} -1 & 0 & 0 & 0 & \dots \\ -2 & -1 & 0 & 0 & \dots \\ -3 & -2 & -1 & 0 & \dots \\ -4 & -3 & -2 & -1 & \dots \\ \vdots & \vdots & \vdots & \vdots & \ddots \end{pmatrix}.$$

It is easy to verify that the endomorphism thus defined is indecomposable, but is not uniformly indecomposable.

We have $\lambda_{\min}(\Lambda) = -1$, $\lambda_{\max}(\Lambda) = 0$. Hence all eigenelements λ of the endomorphism Λ belong to the interval $[-1, 0]$, and an eigenelement of the form $f_\lambda = (0, \lambda, 2\lambda, 3\lambda, \dots)$ corresponds to every number $\lambda \in [-1, 0]$.

Note that the adjoint endomorphism Λ^*, i.e., the endomorphism determined by the matrix

$$\begin{pmatrix} -1 & -2 & -3 & -4 & \dots \\ 0 & -1 & -2 & -3 & \dots \\ 0 & 0 & -1 & -2 & \dots \\ 0 & 0 & 0 & -1 & \dots \\ \vdots & \vdots & \vdots & \vdots & \ddots \end{pmatrix},$$

does not have any eigenelements.

THEOREM 5.1. *Suppose* $\Lambda\colon \mathscr{B}(X, K) \to \mathscr{B}(X, K)$ *is a monotonically continuous uniformly indecomposable endomorphism. Then* Λ *has at most one eigenelement* λ_Λ *which coincides with* $\lambda_{\min}$ ($\lambda_{\min}$ *is the spectral element of* Λ), *if it exists.*

PROOF. We shall show that the eigenelement λ_Λ (if it exists) is a solution of the following "linear programming" problem: to find the minimal value λ for which the set

$$U_\lambda = \left\{ f \in \mathscr{B}(X, K) \mid \Lambda f \leqslant \lambda \odot f,\ \sup_{x\in X} f(x) = \mathbf{1} \right\} \tag{7}$$

is nonempty.

We shall prove that the solution of problem (7) exists and equals λ_Λ. This follows from Assertions 5.1, 5.2 that will be proved below.

ASSERTION 5.1. *Let the set* U_λ *be nonempty for some* λ. *Then* $\lambda_{\min} \leqslant \lambda$.

PROOF. It is clear that it is sufficient to prove the following assertion: for all $p = 0, 1, \dots$ the inequality $\lambda_p \leqslant \lambda$, where $\lambda_p^p = \operatorname{Tr}\Lambda^p$, holds. It is not difficult to verify that for every $p = 0, 1, \dots$ the kernel $K^p(x, y)$ of the endomorphism Λ^p can be expressed in terms of the kernel $K(x, y)$ in the following way:

$$K^p(x, y) = \sup_{x_1, \dots, x_{p-1}} (K(x, x_1) \odot K(x_1, x_2) \odot \dots \odot K(x_{p-1}, y)).$$

Hence to prove Assertion 5.1 it is sufficient to prove the inequality

$$K(x, x_1) \odot K(x_1, x_2) \odot \dots \odot K(x_{p-1}, x) \leqslant \lambda^p, \tag{8}$$

which holds for all points $x_1, x_2, \dots, x_{p-1}$. Let $f \in U_\lambda$. Then for all $z, y \in K$ we have $K(z, y) \odot f(y) \leqslant \lambda \odot f(z)$. Assume that

$$z = x_i, \quad y = x_{i+1} \qquad (i = 0, 1, \dots, p-1), \qquad x_0 = x_p = x.$$

Then from the latter inequality it follows that

$$K(x_i, x_{i+1}) \odot f(x_{i+1}) \leqslant \lambda \odot f(x_i), \qquad i = 0, 1, \dots, p-1.$$

Multiplying these inequalities, we get

$$\begin{aligned} K(x, x_1) \odot \dots \odot K(x_{p-1}, x) \odot f(x) \odot \dots \odot f(x_{p-1}) \\ \leqslant \lambda^p \odot f(x) \odot \dots \odot f(x_{p-1}). \end{aligned}$$

Therefore, taking into account Proposition 4.3 and the cancellation condition, we obtain inequality (8). The proof is over.

ASSERTION 5.2. *The set* $U_{\lambda_{\min}}$ *is nonempty.*

PROOF. Define the endomorphism $\bar{\Lambda}$ in the following way: for a bounded function $f \in \mathscr{B}(X, K)$ put

$$\bar{\Lambda} f(x) = \sup_{p=0,1,\dots} \left(\lambda_{\min}^{-p} \odot \sup_{y \in X} (K^p(x, y) \odot f(y)) \right), \tag{9}$$

where $K^p(x, y)$ is the kernel of the endomorphism Λ^p. Taking into account the expression for $K^p(x, y)$, we obtain the following expression for the kernel $\bar{K}(x, y)$ of the endomorphism $\bar{\Lambda}$:

$$\bar{K}(x, y) = \sup_{p=0,1,\dots} \sup_{x_1, \dots x_{p-1}} (\lambda_{\min}^{-p} \odot K(x, x_1) \odot \cdots \odot K(x_{p-1}, y)).$$

The uniform indecomposability condition implies the estimate

$$\begin{aligned} c \odot \bar{K}(x, y) &\leqslant \sup_{k=0,1,\dots} \sup_{z_1, \dots z_{k-1}} \sup_{p=0,1,\dots} \sup_{x_1, \dots x_{p-1}} (\lambda_{\min}^{-p-k} \odot K(y, z_1) \odot \\ &\qquad \cdots \odot K(z_{k-1}, x) \odot K(x, x_1) \odot \cdots \odot K(x_{p-1}, y)) \\ &\leqslant \sup_{k=0,1,\dots} \sup_{p=0,1,\dots} (\lambda_{\min}^{-p-k} \odot \operatorname{Tr} \Lambda^{p+k}) \leqslant \sup_{s=0,1,\dots} (\lambda_{\min}^{-s} \odot \lambda_s^s) \leqslant \mathbf{1}, \end{aligned}$$

where by definition of $\lambda_{\min}$ we have $\lambda_s^s = \operatorname{Tr} \Lambda^s \leqslant \lambda_{\min}^s$. Thus the inequality $\overline{K}(x, y) \leqslant c^{-1}$ holds. Therefore the endomorphism $\bar{\Lambda}$ has a bounded kernel and so it is a mapping of the semimodule $\mathscr{B}(X, K)$ into itself. Hence for an arbitrary function $f \in \mathscr{B}(X, K)$ we obtain

$$\bar{\Lambda} f \leqslant M \odot \|f\|, \tag{10}$$

where $M = c^{-1}$.

Since (9) implies that for all $f \in \mathscr{B}(X, K)$ we have $\bar{\Lambda} f \geqslant f$, we see that the endomorphism $\bar{\Lambda}$ is nonzero.

Let $f \in \mathscr{B}(X, K)$ be an arbitrary nonzero function. Then the function $g = \bar{\Lambda} f$ is also nonzero. Relations (9) and (10) yield the estimate

$$\begin{aligned} \Lambda g = \Lambda \bar{\Lambda} f &= \Lambda \Big(\sup_{p=0,1,\dots} (\lambda_{\min}^{-p} \odot \Lambda^p f) \Big) = \sup_{p=0,1,\dots} (\lambda_{\min}^{-p} \odot \Lambda^{p+1} f) \\ &= \lambda_{\min} \odot \sup_{p=0,1,\dots} (\lambda_{\min}^{-p-1} \odot \Lambda^{p+1} f) \\ &= \lambda_{\min} \odot \sup_{p=0,1,\dots} (\lambda_{\min}^{-p} \odot \Lambda^p f) \leqslant \lambda \odot g. \end{aligned}$$

Let $\|g\| = \sup_{x \in X} g(x)$. Then the function $\bar{g}(x) = g(x) \odot \|g\|^{-1}$ belongs to the set $U_{\lambda_{\min}}$, and thus, the set $U_{\lambda_{\min}}$ is nonempty.

COROLLARY 5.1. *Let* Λ *be a uniformly indecomposable endomorphism. Then there exists an element* $M \in K$ *such that for all* $f \in \mathscr{B}(X, K)$ *and* $p \in \mathbb{N}$ *we have the inequality*

$$\Lambda^p f \leqslant M \odot \lambda_{\min}^p \odot \|f\|. \tag{11}$$

PROOF. Let f be an arbitrary function from $\mathscr{B}(X, K)$ and $F = \bar{\Lambda} f$, where $\bar{\Lambda}$ is the endomorphism that was constructed previously (see the proof of Assertion 5.2). Then for all p the estimate

$$\Lambda^p f \leqslant \Lambda^p F \leqslant \lambda_{\min}^p \odot F = \lambda_{\min}^p \odot \bar{\Lambda} f$$

is satisfied. Taking into consideration inequality (10) we obtain the required inequality from this estimate.

Now we prove Theorem 5.1. Let λ_Λ be the eigenelement of the endomorphism Λ. Then there exists a nonzero function $f \in \mathscr{B}(X, K)$ such that $\Lambda f \leqslant \lambda_\Lambda \odot f$. Assuming that $\bar{f} = f \odot \|f\|^{-1}$, we obtain $\bar{f} \in U_{\lambda_\Lambda}$ (here U_{λ_Λ} is the set determined by (7)). So the set U_{λ_Λ} is nonempty and then by Assertion 5.1 we get $\lambda_{\min} \leqslant \lambda_\Lambda$. Suppose $\lambda_{\min} < \lambda_\Lambda$. From (11) we get

$$\Lambda^p \bar{f}(x) = \lambda_\Lambda^p \odot \bar{f}(x) \leqslant M \odot \lambda_{\min}^p \odot \|\bar{f}\|$$

for all $x \in X$. Now from the cancellation condition we can conclude that $\lambda_\Lambda^p \leqslant M \odot \lambda_{\min}^p$. But this is equivalent to $M \odot \lambda_{\min}^p = \lambda_\Lambda^p \oplus M \odot \lambda_{\min}^p$, which contradicts the stabilization condition, since $M \neq \mathbf{0}$ and $\lambda_\Lambda^{-1} \odot \lambda_{\min} < \mathbf{1}$. Hence we obtain the desired equality: $\lambda_{\min} = \lambda_\Lambda$. This completes the proof.

As one can see from the next example, there exist uniformly indecomposable endomorphisms that do not possess eigenelements.

EXAMPLE 5.2. Let $K = \mathscr{R}$ (see Example 2.1), $X = \mathbb{N}$, and an endomorphism Λ acting in the semimodule $\mathscr{B}(X, K)$ be uniformly indecomposable. Assume that the infinite matrix

$$\begin{pmatrix} -1 & -1 & -1 & -1 & \dots \\ 0 & -1 & -1 & -1 & \dots \\ 0 & 0 & -1 & -1 & \dots \\ 0 & 0 & 0 & -1 & \dots \\ \vdots & \vdots & \vdots & \vdots & \ddots \end{pmatrix}$$

determines the endomorphism Λ. It can be shown without difficulty that the endomorphism Λ does not have eigenelements.

COROLLARY 5.2. *Let Λ be a uniformly indecomposable endomorphism and $\lambda_{\min}(\Lambda)$, $\lambda_{\max}(\Lambda)$ be its spectral elements. Then $\lambda_{\min}(\Lambda) = \lambda_{\max}(\Lambda)$.*

The proof is straightforward: use Corollary 5.1 and Proposition 3.7.

PROPOSITION 5.1. *Suppose $\Lambda: \mathscr{B}(X, K) \to \mathscr{B}(X, K)$ is a uniformly indecomposable endomorphism and $\lambda_{\min}(\Lambda)$ is its spectral element. Then there exist elements $m, M \in K$ $(m \neq \mathbf{0})$ such that for all $n \in \mathbb{N}$ and $f \in \mathscr{B}(X, K)$ we have the inequality*

$$m \odot \lambda_{\min}^n \odot \inf_{x \in X} f(x) \leqslant \|\Lambda^n f\| \leqslant M \odot \lambda_{\min}^n \odot \|f\|.$$

PROOF. The upper bound

$$\|\Lambda^n f\| \leqslant M \odot \lambda_{\min}^n \odot \|f\|$$

immediately follows from (10). Now we establish the validity of the lower estimate. We have

$$\begin{aligned}\|\Lambda^n f\| &= \sup_{x\in X}\sup_{y\in Y} K^n(x, y)\odot f(y) \geqslant \sup_{x\in X}\sup_{y\in Y} K^n(x, y)\odot \inf_{x\in X} f(x)\\ &= \bar{\lambda}_n^n \odot \inf_{x\in X} f(x) \geqslant \lambda_{\max}^n(\Lambda)\odot \inf_{x\in X} f(x),\end{aligned}$$

whence by Corollary 5.2 we get the required inequality.

COROLLARY 5.3. *Let Λ be a uniformly indecomposable endomorphism. Then the equation $y = \Lambda y \oplus f$ is solvable for any function $f \in \mathcal{B}(X, K)$ if and only if $\lambda_{\min} \leqslant \mathbf{1}$, where the element $\lambda_{\min} \in K$ is the spectral element of the endomorphism Λ.*

PROOF. Let the equation $y = \Lambda y \oplus f$ have a solution for any function $f \in \mathcal{B}(X, K)$. Put $f(x) \equiv 1$. Then the solution of the equation $y = \Lambda y \oplus \mathbf{1}$ is not identically zero. We have the equalities

$$y \oplus \Lambda y = (\Lambda y \oplus \mathbf{1}) \oplus \Lambda y = \Lambda y \oplus \mathbf{1} = y.$$

Therefore $\Lambda y \leqslant y$ and the function $y(x)$ is not identically zero. By letting $\bar{y}(x) = y(x) \odot \|y\|^{-1}$ we obtain a nonzero function satisfying the conditions $\Lambda \bar{y} \leqslant \bar{y}$ and $\sup_{x\in X} \bar{y}(x) = \mathbf{1}$. From Assertion 5.1 it follows that $\lambda_{\min} \leqslant \mathbf{1}$. The necessity of the condition $\lambda_{\min} \leqslant \mathbf{1}$ is proved. Now let us prove its sufficiency. Suppose $\lambda_{\min} \leqslant \mathbf{1}$; then by virtue of Corollary 5.1 the inequality

$$\Lambda^p f \leqslant M \odot \lambda_\Lambda^p \odot \|\Lambda f\| \leqslant M \odot \|f\|$$

is valid for all $f \in \mathcal{B}(X, K)$ and $p \in \mathbb{N}$. Hence for any function $f \in \mathcal{B}(X, K)$ the function $y(x) = \sup_{p=0,1,\dots} (\Lambda^p f)(x)$ belongs to the semimodule $\mathcal{B}(X, K)$ and satisfies the equation $\Lambda y \oplus f = y$. Thus the proof is complete.

REMARK 5.3. Since $\sup(a, b) = a \oplus b$, we see that the solution (found just now) of the equation $\Lambda y \oplus f = y$ can be written as

$$y = \bigoplus_{k=0}^{\alpha} \Lambda^k f;$$

this is the analogue to the "Neumann expansion" in ordinary linear algebra, and the condition $\lambda_\Lambda \leqslant \mathbf{1}$ ensures the convergence of this "expansion". (Note that, unlike the case of ordinary linear algebra, the convergence takes place for $\lambda_\Lambda = \mathbf{1}$.) We see also that in the case when the partial order in the semiring K is not linear, the solution obtained earlier is not unique, even if $\lambda_\Lambda < \mathbf{1}$.

PROPOSITION 5.2. *Suppose $\Lambda: \mathcal{B}(X,K) \to \mathcal{B}(X,K)$ is an arbitrary monotonically continuous uniformly indecomposable endomorphism with kernel $K(x, y)$ and $\lambda_{\min}(\Lambda)$ is its spectral element. Denote by $K^n(x, y)$ the kernel of the endomorphism Λ^n. Assume that there exists a point $x_0 \in X$ such that* *for some number* n_0 *the relation* $\lambda_{\min}(\Lambda) = K^{n_0}(x_0, x_0)$ *is valid. Then the element* $\lambda_{\min}(\Lambda)$ *is the eigenelement for the endomorphism* Λ.

PROOF. First consider the case $n_0 = 1$. Let the function δ_{x_0} be given by

$$\delta_{x_0}(x) = \begin{cases} \mathbf{1} & \text{for } x = x_0, \\ \mathbf{0} & \text{otherwise.} \end{cases}$$

Then for all $x \in X$ the inequality

$$(\Lambda\delta_{x_0})(x) \geqslant \lambda_{\min}(\Lambda) \odot \delta_{x_0}(x) \tag{12}$$

holds. Now we put

$$f = \sup_{p=0,1,\dots} (\lambda_{\min}^{-p}(\Lambda) \odot \Lambda^p \delta_{x_0}).$$

From the proof of Assertion 5.2 and inequality (12) it follows that $f \in \mathscr{B}(X, K)$ and $\Lambda f = \lambda_{\min}(\Lambda) \odot f$, i.e., the function f is the eigenfunction of the endomorphism Λ. Therefore Proposition 5.2 is proved for the case $n_0 = 1$.

Now consider the case $n_0 > 1$. Let $f \in \mathscr{B}(X, K)$ be the eigenfunction of the endomorphism Λ^{n_0} (it exists due to what was proved above). Then the function

$$g = \bigoplus_{i=1}^{n_0} \lambda_{\min}^{n_0 - i}(\Lambda) \odot \Lambda^{i-1} f$$

belongs to the semimodule $\mathscr{B}(X, K)$ and is the eigenfunction of the endomorphism Λ with eigenvalue $\lambda_{\min}(\Lambda)$. And so the proof is complete.

REFERENCES

1. P. I. Dudnikov and S. N. Samborskiĭ, *Endomorphisms of finitely generated free semimodules*, this volume.

CENTER FOR OPTIMIZATION AND MATHEMATICAL MODELING, INSTITUTE OF NEW TECHNOLOGIES, KIROVOGRADSKAYA UL., 11, 113587, MOSCOW, RUSSIA

Translated by I. PETRENKO

ADVANCES IN SOVIET MATHEMATICS
Volume 13, 1992

Endomorphisms of Finitely Generated Free Semimodules

P. I. DUDNIKOV AND S. N. SAMBORSKIĬ

§1. Introduction

A review of the main concepts concerning idempotent semirings, semimodules, and their endomorphisms is given in §§1–4 of this paper; a number of examples is considered. In part these concepts were introduced earlier [1–5]. (In this paper we shall omit the proofs of results obtained in [1–5].) In §5 a result on the existence of an "eigenvalue" and an "eigenvector" of an arbitrary endomorphism of a finitely generated free semimodule is formulated. It has numerous applications in this and other papers constituting this volume. In one particular case (of the semiring $\mathscr{R}$, Example 2.1) this result was obtained earlier in other terms by I. V. Romanovskiĭ [6]. It should be noted that in the case of a metric semiring (which is the case for $\mathscr{R}$) the existence of the "eigenvectors" immediately follows from well-known results of the Brouwer fixed-point theorem type. But the nature of this result is quite different: it is not connected with the topological structures in the semiring but is purely algebraic. In the theorem of §5 the algebraic properties of semirings that allow us to guarantee the existence of "eigenvalues" are distinguished. The maximal eigenvalue, as in classical analysis, gives an upper estimate for iterates of endomorphisms (operators). But in the case of idempotent semirings there is also a lower estimate. A particular case of such an estimate for endomorphisms which have a unique eigenvalue (the so-called indecomposable case) was obtained by I. V. Romanovskiĭ [7].

In the general case, to obtain two-sided estimates, it is necessary to consider the set of all eigenvalues and the order structures on them. This consideration leads to the construction of the spectrum and the spectral function of an endomorphism.

The spectrum of the endomorphism is a graph with weighted vertices. The weights are the eigenvalues of the endomorphism restricted to invariant semimodules. In the case of classical linear operators, the structure of invariant

1991 *Mathematics Subject Classification*. Primary 16Y60.

subspaces is much simpler, and so it is useless to consider any graph structure: it is simply sufficient to know the eigenvalues and their multiplicities. This is not the case for endomorphisms of semimodules, and a corresponding construction is described in §5.

§2. Semirings

DEFINITION 2.1. *A commutative semiring* is defined as a set K with two associative commutative operations $\oplus$ and $\odot$, having neutral elements $\mathbf{0}$ and $\mathbf{1}$, respectively, and satisfying the relations

$$a \odot (b \oplus c) = a \odot b \oplus a \odot c \qquad \text{(distributivity)}$$

and

$$a \odot \mathbf{0} = \mathbf{0}, \quad \forall a \in K.$$

In what follows we shall only consider semirings in which the operation $\oplus$ is *idempotent*, i.e., $a \oplus a = a$ for all $a \in K$.

The idempotent operation $\oplus$ induces a *partial order* $\leqslant$ in the set K according to the rule

$$a \leqslant b \iff a \oplus b = b.$$

Indeed, from the fact that the addition operation $\oplus$ is idempotent, it follows that

(1) $\{a \leqslant b, b \leqslant a\} \Rightarrow a = b$;
(2) $a \leqslant a$, $\forall a \in K$;
(3) $\{a \leqslant b, b \leqslant c\} \Rightarrow a \leqslant c$.

With such a partial order the element $\mathbf{0} \in K$ is the smallest one, therefore all elements of K are "nonnegative".

It is easy to see that the partial order in K thus introduced has the properties:

(1) $a \leqslant b \Rightarrow a \oplus c \leqslant b \oplus c$, $\forall c \in K$;
(2) $a \leqslant b \Rightarrow a \odot c \leqslant b \odot c$, $\forall c \in K$;
(3) $a \oplus b = \sup(a, b)$, $\forall a, b \in K$.

EXAMPLE 2.1. The semiring considered in this example will be very important in the sequel. It will be denoted by $\mathscr{R}$. Consider the set $\mathbb{R}^1 \cup \{-\infty\}$ ($\mathbb{R}^1$ is the real line) and introduce the operations as follows:

$$a \oplus b = \max(a, b), \qquad a \odot b = a + b.$$

Then the set $\mathscr{R} = \mathbb{R}^1 \cup \{-\infty\}$ with the elements $\mathbf{0} = -\infty$ and $\mathbf{1} = 0$ satisfies Definition 2.1. The partial order induced by the operation $\oplus$ is linear and coincides with the usual linear order on $\mathscr{R} \setminus \{\mathbf{0}\} = \mathbb{R}^1$.

EXAMPLE 2.2. Let $K = \mathbb{R}^1_+ \cup \{+\infty\}$, $a \oplus b = \min(a, b)$, $a \odot b = a + b$, $\mathbf{0} = +\infty$, $\mathbf{1} = 0$. In this semiring the partial order induced by the operation $\oplus$ is linear and is opposite to the usual order in K.

EXAMPLE 2.3. Let K_1 and K_2 be commutative semirings with operations $\oplus_i$, $\odot_i$ and neutral elements $\mathbf{0}_i$, $\mathbf{1}_i$ $(i = 1, 2)$, respectively. Then the set $K = K_1 \times K_2$ with the operations $\oplus$ and $\odot$ defined by

$$(a_1, b_1) \oplus (a_2, b_2) = (a_1 \oplus_1 a_2, b_1 \oplus_2 b_2),$$
$$(a_1, b_1) \odot (a_2, b_2) = (a_1 \odot_1 a_2, b_1 \odot_2 b_2),$$

with "zero" $\mathbf{0} = (\mathbf{0}_1, \mathbf{0}_2)$ and "unit" $\mathbf{1} = (\mathbf{1}_1, \mathbf{1}_2)$ is a commutative semiring.

EXAMPLE 2.4. Let K be a semiring and B a system of subsets in K such that B is complete with respect to the following operations:

$$\alpha \oplus \beta = \{x \in K \mid \exists a \in \alpha,\ b \in \beta : x = a \oplus b\},$$
$$\alpha \odot \beta = \{x \in K \mid \exists a \in \alpha,\ b \in \beta : x = a \odot b\}.$$

If the one-point subsets $\{\mathbf{0}\}$ and $\{\mathbf{1}\}$ belong to B, then the set B with the operations introduced above forms a commutative semiring.

EXAMPLE 2.5. If K is the semiring considered in Example 2.1 or Example 2.2 and B is the set of connected intervals of $\mathbb{R}^1 \cup \{-\infty\}$ or of $\mathbb{R}^1_+ \cup \{+\infty\}$, then we have a particular case of Example 2.4.

EXAMPLE 2.6. Suppose K is a collection of compact convex subsets of a convex closed cone in $\mathbb{R}^n$; to any two subsets (elements of K) the operation $\oplus$ assigns their convex hull and the operation $\odot$ their "arithmetic sum", i.e.,

$$A \odot B = A + B = \{a + b \mid a \in A, b \in B\}.$$

Then, assuming that $\mathbf{0} = \varnothing$ (the empty set) and $\mathbf{1} = \{0\}$ (the one-point subset consisting of the vertex 0 of the cone), we get a commutative semiring.

The next condition is very important for the further development of the theory.

CONDITION 2.1 (cancellation). *The relations* $a \odot b = a \odot c$ *and* $a \neq \mathbf{0}$ *imply* $b = c$.

The semirings from Examples 2.1, 2.2, 2.5, and 2.6 satisfy this condition. So do the semirings from Examples 2.3, 2.4 if the cancellation condition holds for the semiring K appearing in those examples.

Let a^n denote the product $a \odot a \odot \cdots \odot a$ of n factors. We state some assertions that will be used in what follows.

ASSERTION 2.1 [4, 5]. *Suppose K is a commutative semiring for which the cancellation condition holds. Then for any elements $a, b \in K$ and any number $n \in \mathbb{N}$ we have the identity*

$$(a \oplus b)^n = a^n \oplus b^n.$$

COROLLARY 2.1. *Let $x_i \in K$ $(i = 1, 2, \dots, n)$. Then for all $p \in \mathbb{N}$ we have the identity*

$$\left(\bigoplus_{i=1}^{n} x_i\right)^p = \bigoplus_{i=1}^{n} x_i^p.$$

COROLLARY 2.2. *Let* $x_i \in K$ $(i = 1, 2, \ldots, n)$. *Then the "Cauchy" inequality*

$$x_1 \odot x_2 \odot \cdots \odot x_n \leqslant x_1^n \oplus x_2^n \oplus \cdots \oplus x_n^n$$

holds. The equality sign appears only if $x_1 = x_2 = \cdots = x_n$.

ASSERTION 2.2. *Suppose the cancellation condition holds in a semiring* K. *If there exists a solution* λ *of the equation* $\lambda^n = a$ *then* λ *is unique.*

The easiest to study are the commutative semirings K in which $\odot$ is a group operation on the set $K \setminus \{\mathbf{0}\}$. If only the cancellation condition holds, then one can extend a semiring to a semiring with group multiplication by using a standard construction of the group of ratios. Because of the additional structure of idempotent addition, we shall carry out this construction in detail.

Let K be a semiring for which the cancellation condition holds. Consider the set of pairs (a, b) $(a, b \in K)$, where $b \neq \mathbf{0}$, with the equivalence relation:

$$(a, b) \sim (c, d), \quad \text{if } a \odot d = b \odot c,$$

pointwise multiplication and the following addition

$$(a, b) \oplus (c, d) = (a \odot d \oplus b \odot c, b \odot d).$$

On the set $\widetilde{K}$ of equivalence classes, one can define a group multiplication, an idempotent addition, a "zero" equal to the coset of $(\mathbf{0}, \mathbf{1})$, a "unit" equal to the coset of $(\mathbf{1}, \mathbf{1})$ (the inverse element to the coset of an element (a, b) with respect to multiplication is the coset of the element (b, a)) and the imbedding $K \to \widetilde{K}$ by the rule $a \mapsto$ {the coset of the element $(a, \mathbf{1})$}. It is easy to check that the distributivity property holds in $\widetilde{K}$, and thus the set $\widetilde{K}$ is a semiring in the sense of Definition 2.1.

If in a semiring K the operation $\odot$ is a group operation (i.e., $K \setminus \{\mathbf{0}\}$ is a group with respect to "multiplication"), then for every pair of elements $a, b \in K$ one can define

$$\inf(a, b) = \begin{cases} (a^{-1} \oplus b^{-1})^{-1} & \text{if } a \odot b \neq \mathbf{0}, \\ \mathbf{0} & \text{if } a \odot b = \mathbf{0}. \end{cases}$$

But, generally speaking, K is not a complete lattice, i.e., infinite bounded subsets do not necessarily possess a supremum and an infimum.

Whenever the semiring K is a complete lattice, the following convergence is defined in K:

$$\{a_n \to a\} \iff \left\{\inf_N \sup_{n>N} a_n = \sup_N \inf_{n>N} a_n = a\right\}. \tag{1}$$

CONDITION 2.2 (stabilization). *A semiring* K *is said to satisfy the stabilization condition, if for all elements* $\lambda, \mu, a, b \in K$ $(a \neq \mathbf{0}, b \neq \mathbf{0})$ *there exist an element* $c \in K$ *and a natural number* $n_0 \in \mathbb{N}$ *such that for every* $n \geqslant n_0$ *we have*

$$a \odot \lambda^n \oplus b \odot \mu^n = c \odot (\lambda \oplus \mu)^n.$$

REMARK 2.1. If a semiring K is such that $K \setminus \{\mathbf{0}\}$ is a group with respect to the multiplication $\odot$, then the stabilization condition is equivalent to the following condition: for all elements $\lambda, b \in K$ such that $\lambda \leqslant \mathbf{1}$ and $b \neq \mathbf{0}$ the sequence $\{\lambda^n \oplus b\}_{n=0}^{\infty}$ is stabilized.

ASSERTION 2.3. *Suppose that a semiring K satisfies the stabilization condition and the set $K \setminus \{\mathbf{0}\}$ is a group with respect to the operation $\odot$. Then whenever the operation $\oplus$ is continuous with respect to the topology defined by* (1), *the partial order in K induced by $\oplus$ is linear.*

PROOF. Let $a, b \in K$ be arbitrary elements. We shall show that either $a \leqslant b$ or $b \leqslant a$. Let $c = a \oplus b$. It is clear that one can assume that $a \neq \mathbf{0}$ and $b \neq \mathbf{0}$. Suppose $c \neq a$ (otherwise the equality $a = a \oplus b$, which is equivalent to the inequality $b \leqslant a$, would hold). Let $a_1 = a \odot c^{-1}$, $b_1 = b \odot c^{-1}$. Then $a_1 \oplus b_1 = \mathbf{1}$, where $a_1 \leqslant \mathbf{1}$ and $a_1 \neq \mathbf{1}$. Now we prove that $a_1^n \oplus b_1 = \mathbf{1}$ for all $n \geqslant 1$.

We argue by induction. For $n = 1$ the assertion is evident. Now suppose $a_1^k \oplus b_1 = \mathbf{1}$. Then let us prove that $a_1^{k+1} \oplus b_1 = \mathbf{1}$. We have

$$\begin{aligned}\mathbf{1} = a_1^k \oplus b_1 &= (a_1^k \oplus b_1) \odot (a_1 \oplus b_1) = a_1^{k+1} \oplus a_1 \odot b_1 \oplus a_1^k \odot b_1 \oplus b_1^2 \\ &= a_1^{k+1} \oplus a_1 \odot b_1 \oplus b_1 \odot (a_1^k \oplus b_1) = a_1^{k+1} \oplus a_1 \odot b_1 \oplus b_1 \\ &= a_1^{k+1} \oplus b_1 \odot (a_1 \oplus \mathbf{1}) = a_1^{k+1} \oplus b_1.\end{aligned}$$

Hence $a_1^n \oplus b_1 = \mathbf{1}$ holds for all $n \geqslant 1$.

From the stabilization condition it follows that whenever $a_1 \leqslant \mathbf{1}$ and $a_1 \neq \mathbf{1}$ the limit $\lim_{n\to\infty} a_1^n$ exists and is equal to $\mathbf{0}$. Thus we have

$$\mathbf{1} = \lim_{n\to\infty} (a_1^n \oplus b_1) = \lim_{n\to\infty} a_1^n \oplus b_1 = b_1.$$

This yields $b \odot c^{-1} = \mathbf{1}$, which is equivalent to $b = c = a \oplus b$. This means that $a \leqslant b$ and Assertion 2.3 is proved.

CONDITION 2.3 (algebraic completeness). *A semiring K is called algebraically complete, if the equation $x^n = x \odot x \odot \cdots \odot x = a$ has a solution x for an arbitrary $a \in K$ and any positive integer n.*

REMARK 2.2. If a semiring K satisfies the cancellation condition and is algebraically complete then Assertion 2.2 implies that the equation $x^n = a$ has a unique solution.

EXAMPLE 2.7. Let K_i $(i = 1, 2, \dots, n)$ be semirings which satisfy the cancellation condition (Condition 2.1), the stabilization condition (Condition 2.2), and the condition of algebraic completeness (Condition 2.3). Consider the following semiring

$$K = (K_1 \setminus \{\mathbf{0}\}) \times (K_2 \setminus \{\mathbf{0}\}) \times \cdots \times (K_n \setminus \{\mathbf{0}\}) \cup (\mathbf{0}, \dots, \mathbf{0})$$

with operations

$$a \oplus b = (a_1 \oplus b_1, a_2 \oplus b_2, \dots, a_n \oplus b_n),$$
$$a \odot b = (a_1 \odot b_1, a_2 \odot b_2, \dots, a_n \odot b_n),$$

where $a = (a_1, a_2, \dots, a_n)$, $b = (b_1, b_2, \dots, b_n)$ are elements from K.

It can be directly verified that the semiring K satisfies the cancellation, stabilization, and algebraic completeness conditions. In particular, the semiring $\overline{\mathbb{R}}^n = \mathbb{R}^n \cup (-\infty, -\infty, \dots, -\infty)$ in which the operation $\oplus$ induces a partial order (which is not linear) satisfies these conditions.

§3. Semimodules

DEFINITION 3.1. A set M is called a *semimodule over the semiring* K, if there is a commutative associative addition operation $\oplus$ in M with neutral element $\mathbf{0}$ and a multiplication $\odot$ of elements from M by elements of the semiring K is defined, these operations satisfying the following conditions:

$$(\lambda \odot \mu) \odot m = \lambda \odot (\mu \odot m) \qquad \text{for all } \lambda, \mu \in K, \ m \in M,$$
$$\lambda \odot (m_1 \oplus m_2) = \lambda \odot m_1 \oplus \lambda \odot m_2 \qquad \text{for all } \lambda \in K, \ m_1, m_2 \in M,$$
$$\mathbf{0} \odot m = \lambda \odot \mathbf{0} = \mathbf{0} \qquad \text{for all } \lambda \in K, \ m \in M,$$
$$\text{if } \sup_\alpha \lambda_\alpha \in K \text{ and } m \in M \text{ then } \left(\sup_\alpha \lambda_\alpha\right) \odot m = \sup_\alpha(\lambda_\alpha \odot m).$$

Everywhere below the "addition" $\oplus$ is assumed to be idempotent, i.e., $m \oplus m = m$ for all $m \in M$.

EXAMPLE 3.1. Let X be some set and K a semiring. In K the operation $\oplus$ is idempotent (this assumption was made at the beginning of the paper), so K is endowed with a partial order $\leqslant$. Denote by $\mathscr{B}(X, K)$ the set of bounded (in the sense of the partial order in K) functions on X with values in K, i.e. $f \in \mathscr{B}(X, K)$ if and only if there exists an element $\mu \in K$ such that $f(x) \leqslant \mu$ for all $x \in X$. The set $\mathscr{B}(X, K)$ is a semimodule over the semiring K with respect to the operations

$$(f \oplus g)(x) = f(x) \oplus g(x), \qquad f, g \in \mathscr{B}(X, K), \ x \in X,$$
$$(\lambda \odot f)(x) = \lambda \odot f(x), \qquad f \in \mathscr{B}(X, K), \ x \in X, \ \lambda \in K.$$

If the set K is finite and consists of n elements, then the corresponding semimodule $\mathscr{B}(X, K)$ is the collection of "vectors" $(a_1, a_2, \dots, a_n)$ $(a_i \in K)$ and is denoted by K^n.

EXAMPLE 3.2. Let X be a topological space and $K = \mathscr{R}$ (see Example 2.1). Then the set $C(X, \mathscr{R})$ of continuous functions on X with values in $\mathscr{R}$ is a semimodule over the semiring $\mathscr{R}$. If X is a compact set, then the semimodule $C(X, \mathscr{R})$ is a subsemimodule (in the obvious sense) of the semimodule $\mathscr{B}(X, K)$.

EXAMPLE 3.3. Let $X = \mathbb{R}^n$ and $K = \mathscr{R}$. Then the set $\mathrm{Conv}(X, \mathscr{R})$ of concave functions on X with values in $\mathscr{R}$ is a semimodule over the semiring $\mathscr{R}$.

§4. Free finitely generated semimodules

Suppose K is a commutative semiring with an idempotent addition $\oplus$. The semimodule structure in $K^n = K \times K \times \cdots \times K$ over the semiring K is defined by the addition of "vectors"

$$\begin{aligned} a \oplus b &= (a_1, a_2, \dots, a_n) \oplus (b_1, b_2, \dots, b_n) \\ &= (a_1 \oplus b_1, a_2 \oplus b_2, \dots, a_n \oplus b_n) \end{aligned}$$

and multiplication by elements from K ("scalars")

$$\lambda \odot a = \lambda \odot (a_1, a_2, \dots, a_n) = (\lambda \odot a_1, \lambda \odot a_2, \dots, \lambda \odot a_n).$$

The system of n elements of the form

$$e_i = (\mathbf{0}, \mathbf{0}, \dots, \mathbf{0}, \mathbf{1}, \mathbf{0}, \dots, \mathbf{0})$$

is a *system of generators for the semimodule* K^n. This means that any element $a \in K^n$ can be written in the form

$$a = a_1 \odot e_1 \oplus a_2 \odot e_2 \oplus \cdots \oplus a_n \odot e_n.$$

ASSERTION 4.1 [4, 5]. *In K^n any system of n generators consists of the elements of the form $(\mathbf{0}, \mathbf{0}, \dots, \mathbf{0}, \delta_i, \mathbf{0}, \dots, \mathbf{0})$, where $\delta_i \neq \mathbf{0}$.*

REMARK 4.1. The minimal number of generators in the semimodule K^n is equal to n, but for $n > 1$ semimodules in which the minimal system of generators contains as many elements as one wants can exist in K^n, even if $\odot$ is a group operation. For instance, in $\mathscr{R}^3$ ($\mathscr{R}$ is the semiring in Example 2.1) consider the subsemimodule generated by elements of the set Λ, where $\Lambda \subset \mathbb{R}^3$ is a two-dimensional disk of radius 1 which is orthogonal to the vector $(1, 1, 1) \in \mathbb{R}^3$. This subsemimodule on $\mathscr{R}^3 \setminus \{\mathbf{0}\}$ (the set $\mathscr{R}^3 \setminus \{\mathbf{0}\}$ is identified with $\mathbb{R}^3$) is a circular cylinder with axis directed along the vector $(1, 1, 1)$, and there does not exist a finite system of generators that span this subsemimodule.

DEFINITION 4.1. Any system of n generators in K^n will be called a *basis* in K^n.

ASSERTION 4.2. *Suppose X is a set, K is a commutative semiring with idempotent operation $\oplus$, and the semimodule $\mathscr{B}(X, K) = V_1 \dotplus V_2$, where $\dotplus$ is the direct sum (i.e., every element φ can be uniquely expressed in the form $\varphi = v_1 \oplus v_2$); then there exists a decomposition $X = X_1 \cup X_2$, $X_1 \cap X_2 = \varnothing$ such that $V_i = \mathscr{B}(X_i, K)$, and $v_i \in \mathscr{B}(X_i, K)$ is extended to X by zero outside X_i.*

PROOF. It is easy to see that $\mathscr{B}(X, K) = V_1 \dotplus V_2$ implies $V_1 \cap V_2 = \{\mathbf{0}\}$, where $\mathbf{0}$ is the neutral element of the semimodule $\mathscr{B}(X, K)$. Let $\{\delta_u(x) \mid u \in U\}$ be a basis of the semimodule $\mathscr{B}(X, K)$, i.e.,

$$\delta_u(x) = \begin{cases} \mathbf{1} & \text{for } x = u, \\ \mathbf{0} & \text{otherwise.} \end{cases}$$

We shall show that every function $\delta_u(x)$ belongs to one of the semimodules V_i $(i = 1, 2)$. Since $\mathscr{B}(X, K) = V_1 \dotplus V_2$, we see that for every $u \in U$ there exist elements $f_i(x) \in V_i$ $(i = 1, 2)$ such that $\delta_u(x) = f_1(x) \oplus f_2(x)$. For $x \neq u$ the relation $\delta_u(x) = 0$ implies $f_1(x) \oplus f_2(x) = \mathbf{0}$ which in turn yields $f_i(x) = \mathbf{0}$ $(i = 1, 2)$. Therefore $f_i(x) = a_i \odot \delta_u(x)$ $(i = 1, 2)$ for some $a_i \in K$. We have

$$\mathbf{1} = \delta_u(u) = f_1(u) \oplus f_2(u) = a_1 \odot \delta_u(u) \oplus a_2 \odot \delta_u(u) = a_1 \oplus a_2,$$

so at least one of the elements a_i is not $\mathbf{0}$. Let that element be a_1. Then $\delta_u(x) = a_1^{-1} \odot f_1(x) \in V_1$ and since the choice $\delta_u(x)$ is arbitrary, every function $\delta_u(x)$ belongs to one of the semimodules V_i $(i = 1, 2)$.

Let $X_i = \{u \in X \mid \delta_u(x) \in V_i\}$. From the above it follows that $X = X_1 \cup X_2$, $X_1 \cap X_2 = \varnothing$ and $\mathscr{B}(X_i, K) \subseteq V_i$. Now we shall prove the converse inclusion $V_i \subseteq \mathscr{B}(X_i, K)$. Let $f(x) \in V_1$ (for $i = 2$ the proof is the same). Suppose $f_i = f|_{X_i}$ is the restriction of the function f to the set X_i $(i = 1, 2)$. Then

$$f_i(x) = \sup_{u \in X_i} f_i(u) \odot \delta_u(x) \in \mathscr{B}(X_i, K) \subseteq V_i$$

and $f = f_1 \oplus f_2$. Since the expression of any element from the set $\mathscr{B}(X, K)$ is unique, we obtain $f_1(x) = f(x)$ and $f_2(x) \equiv \mathbf{0}$ for all $x \in X$. Hence $f(x) \in \mathscr{B}(X_1, K)$, and thus, $V_1 \subseteq \mathscr{B}(X_1, K)$. Assertion 4.2 is proved.

§5. Endomorphisms

Assume that V_1 and V_2 are semimodules over a semiring K. Then the mapping $A: V_1 \to V_2$ is a *homomorphism of semimodules*, if for all $f, g \in V_1$, $\lambda, \mu \in K$ we have

$$A(\lambda \odot f \oplus \mu \odot g) = \lambda \odot Af \oplus \mu \odot Ag.$$

Homomorphisms from V to V are called *endomorphisms*. In the sequel, the set of endomorphisms of V will be denoted by $\operatorname{End}(V)$.

If V is the finitely generated free semimodule over the semiring K (i.e., if $V = K^n$), then in any basis $\{e_i\}$ in K^n an endomorphism $A \in \operatorname{End}(V)$ can be written by means of a matrix (a_{ij}) as follows:

$$(Af)_i = \bigoplus_{j=1}^{n} a_{ij} \odot f_j.$$

In the case when $V = \mathscr{B}(X, K)$ any $A \in \operatorname{End}(V)$ is an "integral operator with kernel a", i.e., $(Af)(x) = \sup_{y \in X}(a(x, y) \odot f(y))$.

An important source of endomorphisms are problems of dynamic programming. Consider the simplest one of them.

Suppose we are given two sets (the *state space* X and the *control space* Y) and a mapping $f\colon X \times Y \to X$ assigning to each point $x \in X$ and to each control $y \in Y$ the new point $f(x, y) \in X$ that can be "reached" from x by use of the control y. The choice of the initial state $x_0 \in X$ and the collection of controls $\{y_1, y_2, \dots, y_n\}$ determines an *admissible trajectory* $(x_0, x_1, \dots, x_n)$ in X, where $x_i = f(x_{i-1}, y_i)$.

Let also a semiring K and a "multiplicative" K-valued criterion Φ on the set of admissible trajectories be given. This means that the mapping Φ assigns to every collection

$$(x_0, x_1, \dots, x_{k-1}), (y_1, y_2, \dots, y_k) \qquad (x_i \in X, \ y_i \in Y)$$

an element from K of the form

$$\varphi(x_0, y_1) \odot \varphi(x_1, y_2) \odot \cdots \odot \varphi(x_{k-1}, y_k),$$

where $\varphi\colon X \times Y \to K$.

The classical situation corresponds to the case $K = \mathscr{R}$. Then

$$\Phi(x_0, x_1, \dots, x_{k-1}, y_1, \dots, y_k) = \sum_{i=0}^{k-1} \varphi(x_i, y_{i+1}). \tag{2}$$

Let us consider the problem of finding all k-step "extremal" trajectories $(x_0^*, x_1^*, \dots, x_k^*)$, $x_i^* = f(x_{i-1}^*, y_i^*)$, i.e., trajectories such that

$$\begin{aligned} &\Phi(x_0^*, x_1^*, \dots, x_{k-1}^*, y_1^*, \dots, y_k^*) \\ &\qquad \geqslant \Phi(x_0, x_1, \dots, x_{k-1}, y_1, \dots, y_k), \end{aligned} \tag{3}$$

where on the right-hand side of (3) all possible k-step admissible trajectories $(x_0, x_1, \dots, x_{k-1})$ $(x_i = f(x_{i-1}, y_i))$ that start from the point x_0 can appear. (The order structure in K is induced by the idempotent addition.)

To understand the concept of "being extremal", let us consider some specific situations. Let $K = \mathscr{R}$ or, in a more general case, let the partial order in K be linear (see Examples 2.2, 2.3). This generalization will not produce any changes in the sequel. Then inequality (3) must be understood "literally":

$$\Phi(x_0^*, \dots, y_k^*) = \sup \Phi(x_0, \dots, x_{k-1}, y_1, \dots, y_k),$$

where sup (in $\mathscr{R}$) is taken over all admissible k-step trajectories beginning at the point x_0.

For every $k \in \mathbb{N}$ let us introduce the *Bellman function* $\omega_k(x)$ corresponding to the problem of dynamic programming: $\omega_k(x)$ is the maximal value of the criterion Φ on the set of all admissible k-step trajectories that start from the point x_0. Using the specific form of the criterion (its "additivity" for $K = \mathscr{R}$), we obtain from (2) the following recursive relation:

$$\omega_{k+1}(x) = \sup_{y \in Y} \{\varphi(x, y) + \omega_k(f(x, y))\}.$$

Now consider the semimodule $\mathscr{B}(X, \mathscr{R})$. Let $A: \mathscr{B}(X, \mathscr{R}) \to \mathscr{B}(X, \mathscr{R})$ be the mapping given by

$$(A\omega)(x) = \sup_{y \in Y}\{\varphi(x, y) + \omega(f(x, y))\},$$

where $\omega \in \mathscr{B}(X, \mathscr{R})$.

ASSERTION 5.1. *For* $\varphi \in \mathscr{B}(X \times Y, \mathscr{R})$ *the mapping* A *is an endomorphism of the semimodule* $\mathscr{B}(X, \mathscr{R})$.

This assertion is proved by a direct verification.

The Bellman function $\omega_k(x)$ can now be written in the form

$$\omega_k(x) = (A^n \omega_0)(x),$$

where $\omega_0(x) \equiv \mathbf{1}$.

§6. Proper subsemimodules

Let K be a semiring satisfying the algebraic completeness condition, and A an endomorphism of the semimodule K^n, i.e. a mapping from K^n into K^n such that for all $\lambda, \mu \in K$, $f, g \in K^n$ we have

$$A(\lambda \odot f \oplus \mu \odot g) = \lambda \odot Af \oplus \mu \odot Ag.$$

In a basis $\{e_i\}_{i=1}^n$ in K^n the endomorphism A can be written using the matrix (a_{ij}) as follows:

$$(Af)_i = \bigoplus_{j=1}^{n} a_{ij} \odot f_j.$$

An endomorphism $A \in \operatorname{End}(K^n)$ is said to be *decomposable*, if there exists a basis in which A has the form

$$\begin{bmatrix} A_{11} & A_{12} \\ \mathbf{0} & A_{22} \end{bmatrix},$$

where A_{11}, A_{22} are square matrices and $\mathbf{0}$ is the null matrix. By virtue of Assertion 4.1, to check whether an endomorphism $A \in \operatorname{End}(K^n)$ is decomposable we must simultaneously transpose rows and corresponding columns in the matrix of A written in any basis, in particular, in the standard one. An endomorphism which is not decomposable will be called *indecomposable*. The following assertion is true:

ASSERTION 6.1. *The matrix of an endomorphism* $A \in \operatorname{End}(K^n)$ $(n > 1)$ *is indecomposable iff for any indices* i *and* j *there exists a sequence of indices* $j_1, j_2, \ldots, j_s$ *such that*

$$a_{ij_1} \odot a_{j_1 j_2} \odot \cdots \odot a_{j_s j} \neq \mathbf{0}.$$

Let the matrix (a_{ij}) correspond to an endomorphism A in the standard basis in K^n. Assume that the matrix (a_{ij}) is indecomposable. Let $\varphi(n)$ be the least common multiple of the numbers $1, \dots, n$ and λ be the root of the equation

$$\lambda^{\varphi(n)} = \bigoplus_{\substack{k=1,2,\dots,n \\ (i_1, i_2, \dots, i_k)}} [a_{i_1 i_2} \odot a_{i_2 i_3} \odot \cdots \odot a_{i_k i_1}]^{\varphi(n)/k}, \tag{4}$$

where the summation is over k and, for each k, over all k-tuples of indices $(i_1, i_2, \dots, i_k)$. The existence and the uniqueness of this root follow from the algebraic completeness condition of the semiring K and Assertion 2.2. The fact that the matrix (a_{ij}) is indecomposable and Assertion 6.1 imply $\lambda \neq \mathbf{0}$ for $n > 1$.

For instance, equation (4) has the following form in K^2 and K^3 respectively:

$$(1) \quad \lambda^2 = a_{11}^2 \oplus a_{22}^2 \oplus a_{12} \odot a_{21};$$

$$(2) \quad \lambda^3 = a_{11}^2 \oplus a_{22}^2 \oplus a_{33}^2 \oplus (a_{12} \odot a_{21})^3 \oplus (a_{13} \odot a_{31})^3 \oplus (a_{23} \odot a_{32})^3 \oplus (a_{12} \odot a_{23} \odot a_{31})^2 \oplus (a_{13} \odot a_{32} \odot a_{21})^2.$$

In particular, for the semiring $\mathscr{R}$ when $n = 2$ we obtain

$$\lambda = \max[a_{11}, a_{22}, (a_{12} + a_{21})/2].$$

THEOREM 6.1 [3–5]. *Let the semiring K satisfy the cancellation, stabilization, and algebraic completeness conditions, A be an indecomposable endomorphism acting in K^n, and λ be the root of equation* (4). *Then there exist natural numbers m_1 and m_2 such that $m_1 > m_2$ and*

$$A^{m_1} = \lambda^{m_1 - m_2} \odot A^{m_2}. \tag{5}$$

THEOREM 6.2 [3–5]. *Let a semiring K satisfy the cancellation, stabilization, and algebraic completeness conditions. Then for every endomorphism $A \in \mathrm{End}(K^n)$ there exist a subsemimodule $J \subset K^n$ $(J \neq \{\mathbf{0}\})$ (an "eigensubspace") and an element $\lambda \in K$ (an "eigenvalue") such that*

$$Ax = \lambda \odot x \quad \text{for all } x \in J.$$

If the endomorphism A is indecomposable, this element λ is unique.

§7. Spectra and estimates of iterates of endomorphisms

ASSERTION 7.1. *Suppose K is a semiring satisfying the cancellation, stabilization, and algebraic completeness conditions, the set $K \setminus \{\mathbf{0}\}$ being a group under the multiplication $\odot$. Further let A be an indecomposable endomorphism acting in K^n with the eigenvalue λ. Then for any element $x = (x_1, x_2, \dots, x_n) \in K^n$, where $x_i > \mathbf{0}$ $(i = 1, 2, \dots, n)$, there exist* *elements* $m_x, M_x \in K$ $(m_x \neq \mathbf{0})$ *such that for all* $k \geqslant 0$ *the estimate*

$$m_x \odot \lambda^k \odot x \leqslant A^k x \leqslant M_x \odot \lambda^k \odot x \tag{6}$$

is valid.

PROOF. Let k_1 and k_2 be numbers such that $A^{k_2} = \lambda^{k_2-k_1} \odot A^{k_1}$. Without loss of generality we can assume that $\lambda = \mathbf{1}$. (Whenever $\lambda = \mathbf{0}$, estimate (6) is obviously true.) Then for all $p \geqslant k_2$ the relations

$$A^p = A^{k_1+(p-k_2)\mathrm{mod}(k_2-k_1)}$$

hold, and therefore, it is sufficient to prove estimate (6) for the case $k \leqslant k_2$. Let $x^0 = x$, $x^{i+1} = Ax^i$, $i = 0, 1, 2, \dots, k_2 - 1$. Then it is clear that the inclusion

$$A^k x \in \{x^0, x^1, \dots, x^{k_2-1}\}$$

holds. Since $x_i > \mathbf{0}$ for all $i = 1, 2, \dots, n$ and A is an indecomposable endomorphism, we have $(A^k x)_i > \mathbf{0}$ for all $i = 1, 2, \dots, n$; $k = 0, 1, 2, \dots, k_2 - 1$. This inequality implies $x_i^j > \mathbf{0}$ for all $i = 1, 2, \dots, n$; $j = 0, 1, 2, \dots, k_2 - 1$, and so the element $m = \inf_{i,j} x_i^j$ is not equal to the neutral element in K. Suppose also $M = \sup_{i,j} x_i^j$. Then for all $i = 1, 2, \dots, n$; $j = 0, 1, \dots, k_2 - 1$ the inequalities $m \leqslant x_i^j \leqslant M$ hold and yield the estimate $m \leqslant (A^k x)_i \leqslant M$, which is true for all $k = 0, 1, 2, \dots$. Let $\alpha = \inf_i x_i$, $\beta = \sup_i x_i$. Then for all $i = 1, 2, \dots, n$ the inequalities $\alpha \leqslant x_i \leqslant \beta$ are satisfied. Since $x_i > \mathbf{0}$ for all $i = 1, 2, \dots, n$, we obtain $\alpha > \mathbf{0}$. Thus we have

$$m \odot \beta^{-1} \odot x_i \leqslant m \odot \beta^{-1} \odot \beta \leqslant (A^k x)_i \leqslant M \odot \alpha^{-1} \odot \alpha \leqslant M \odot \alpha^{-1} \odot x_i.$$

Assuming that $m_x = m \odot \beta^{-1}$, $M_x = M \odot \alpha^{-1}$, we obtain the required estimate.

Now we consider the general case of a decomposable endomorphism. Denote by X the set $\{1, 2, \dots, n\}$. Let A be an arbitrary endomorphism from $\mathrm{End}(K^n)$ defined by the matrix $\widetilde{A} = (a_{ij})_{i,j=1}^r$ (in the standard basis in K^n) and A^S be the endomorphism defined by the matrix $\widetilde{A} \oplus \widetilde{A}'$, where $\widetilde{A}' = (a_{ji})_{i,j=1}^r$ is the transposed matrix.

DEFINITION 7.1. An endomorphism A acting in K^n is called *irreducible*, if the corresponding endomorphism A^S is indecomposable.

ASSERTION 7.2. *An endomorphism* A *acting in* K^n *can be expressed in the following form*:

$$A = \dotplus_{i=1}^{n_1} A_i,$$

where $\dotplus$ *is the direct sum,* A_i $(i = 1, 2, \dots, n_1)$ *is an irreducible endomorphism. In other words, the standard basis in* K^n *can be renumbered so that the matrix* $\widetilde{A}$ *of the endomorphism* A *has the following form in the renumbered basis*:

$$\widetilde{A} = \begin{bmatrix} \widetilde{A}_1 & \mathbf{0} & \dots & \mathbf{0} \\ \mathbf{0} & \widetilde{A}_2 & \dots & \mathbf{0} \\ \vdots & \vdots & \ddots & \vdots \\ \mathbf{0} & \mathbf{0} & \dots & \widetilde{A}_{n_1} \end{bmatrix},$$

where $\widetilde{A}_i$ $(i = 1, 2, \dots, n_1)$ *is the matrix of an irreducible endomorphism in the corresponding semimodule.*

To every endomorphism A we assign a *decomposition* $X = \bigcup_{I=1}^{m} X_I$ of the finite set $X = \{1, \dots, n\}$ and an *oriented graph* $\Gamma(A)$ with m vertices (corresponding to the m subsets $X_1, \dots, X_m$ in the decomposition $X = \bigcup_{I=1}^{m} X_I$) by the rule:

(1) (*Definition of the decomposition*). On the set X we introduce the equivalence relation $\sim$ as follows: $i \sim j$, if one of the following conditions holds:

i) $i = j$;

ii) there exist numbers p and q and sequences of indices $i_1, i_2, \dots, i_p$ and $j_1, j_2, \dots, j_q$ such that

$$a_{ii_1} \odot a_{i_1 i_2} \odot \dots \odot a_{i_p j} > \mathbf{0} \quad \text{and} \quad a_{jj_1} \odot a_{j_1 j_2} \odot \dots \odot a_{j_q i} > \mathbf{0},$$

where a_{sk} are elements of the matrix $\widetilde{A}$ corresponding to the endomorphism A (in the standard basis in K^n).

(2) (*Construction of the graph* $\Gamma(A)$). The sets X_I stand for vertices of the graph $\Gamma(A)$. The vertices X_I and X_J are joined by an edge directed from X_I to X_J if and only if there exist indices $i_0 \in X_I$ and $j_0 \in X_J$ such that $a_{i_0 j_0} > \mathbf{0}$. If X_I and X_J are joined by an edge directed from X_I to X_J, then these vertices are not joined by an edge directed from X_J to X_I $(I \neq J)$.

Denote by A_i the endomorphism acting in $K^{|X_I|}$ ($|X_I|$ is a number of elements of the set X_I) which is determined by the matrix $(a_{sj})_{s, j \in X_I}$.

ASSERTION 7.3. *For all* X_I, *the endomorphism* A_I *is either equal to zero or is indecomposable.*

To every vertex X_I we assign the eigenvalue λ_I of the endomorphism A_I.

Let X_I and X_J be vertices of the graph $\Gamma(A)$. We shall say that there exists a *path starting from* X_I *and ending in* X_J, if there exists a sequence of vertices $X_{k_1}, X_{k_2}, \dots, X_{k_p}$ such that the vertices X_I and X_{k_1} are joined by an edge directed from X_I to X_{k_1}, for all $s = 1, 2, \dots, p$ the vertices X_{k_s} and $X_{k_{s+1}}$ are joined by an edge directed from X_{k_s} to $X_{k_{s+1}}$, and, finally, the vertices X_{k_p} and X_J are joined by an edge directed from X_{k_p} to X_J.

PROPOSITION 7.1. *For the existence of a path starting from X_I and ending in X_J, it is necessary and sufficient that for all indices $i \in X_I$ and $j \in X_J$ one of the following conditions holds*:

i) $a_{ij} > \mathbf{0}$;
ii) *there exists a natural number s and a sequence of indices $k_1, \dots, k_s$ such that $a_{ik_1} \odot a_{k_1k_2} \odot \cdots \odot a_{k_sj} > \mathbf{0}$.*

DEFINITION 7.2. Let X' be a subset of vertices of the graph $\Gamma(A)$. The *overgraph* Γ' of the set X' is the subgraph of the graph $\Gamma(A)$ defined as follows:

i) X_I is a vertex of Γ' iff either $X_I \in X'$ or there exists a vertex $X_J \in X'$ such that there exists a path starting out from X_I and ending in X_J;
ii) two vertices X_I and X_J of Γ' are joined by an edge directed from X_I to X_J iff they are joined by such an edge as vertices of the graph $\Gamma(A)$.

Let Y be some subset of the set X and $\Gamma_Y(A)$ be the overgraph of the set $\{X_I \in \Gamma(A) \mid X_I \cap Y \neq \varnothing\}$.

We denote by $\Gamma_{X_I, Y}(A)$ the subgraph of $\Gamma_Y(A)$ formed by the vertex X_I and the vertices that can be reached by paths from X_I. Vertices are joined by an edge in $\Gamma_{X_I, Y}(A)$, if they are joined by an edge in $\Gamma_Y(A)$.

DEFINITION 7.3. The *spectral function* $\lambda_Y(\cdot): X \to K$ *of the endomorphism* $A \in \mathrm{End}(K^n)$ *corresponding to a subset* Y *of the form*

$$\bigcup_{\{I \mid X_I \cap Y = \varnothing\}} X_I$$

is defined to be the mapping

$$\lambda_Y(x) = \begin{cases} \mathbf{0} & \text{if } X_I \text{ is not a vertex of } \Gamma_Y(A), \\ \sup\{\lambda_J \mid X_J \in \Gamma_{X_I, Y}(A)\} & \text{if } X_I \text{ is a vertex of } \Gamma_Y(A). \end{cases}$$

THEOREM 7.1. *Let A be an arbitrary endomorphism acting in K^n, $X = \{1, 2, \dots, n\}$, and $X = \bigcup_{I=1}^m X_I$ be the decomposition of X corresponding to A. If $f \in K_s^n$ satisfies $\mathrm{supp}(A^{s-1} f) = \mathrm{supp}(A^s f)$ for some $s \in \mathbb{N}$, then there exist elements $m_f, M_f \in K^n$ $(m_f \neq \mathbf{0})$ and a natural number $k_0 \in \mathbb{N}$ such that for all $k \geqslant k_0$ we have the estimate*

$$m_f \odot \lambda_Y^k(i) \leqslant (A^k f)_i \leqslant M_f \odot \lambda_Y^k(i), \tag{7}$$

where $Y = \mathrm{supp}\, f$ and $\lambda_Y(i)$ is the spectral function of A.

The proof immediately follows from Assertions 7.4.1–7.4.7 that will be proved below.

ASSERTION 7.4.1. *Let* $f \in K^n$, $Y = \operatorname{supp} f$. *Denote* $\widetilde{Y} = \bigcup_{X_I \in \Gamma_Y(A)} X_I$. *Then*

$$\widetilde{Y} = \bigcup_k \operatorname{supp} A^k f.$$

PROOF. Let $i \in \operatorname{supp} A^k f$ for some k. If $k = 0$ then the inclusion $i \in \widetilde{Y}$ is obvious. Suppose $k > 1$. (For $k = 1$ the proof of the inclusion $i \in \widetilde{Y}$ is analogous to what follows.) Then

$$\bigoplus_{j=1}^{n} a_{ij}^{(k)} \odot f_j > \mathbf{0}$$

(here $(a_{ij}^{(k)})_{i,j=1}^n$ is the matrix of the endomorphism A^k in the standard basis). Hence $a_{ij}^{(k)} > \mathbf{0}$ for some $j \in Y$. The inequality $a_{ij}^{(k)} > \mathbf{0}$ means that there exists a sequence of indices $i_1, i_2, \dots, i_{k-1}$ such that

$$a_{ii_1} \odot a_{i_1 i_2} \odot \cdots \odot a_{i_{k-1} j} > \mathbf{0}.$$

Suppose the indices i and j determine the vertices X_α and X_β of the graph $\Gamma(A)$ respectively. The last inequality is equivalent to the existence in $\Gamma(A)$ of a path starting from the vertex X_α and ending at the vertex X_β. Since $j \in Y$, we obtain $X_\beta \cap Y \neq \varnothing$, and therefore, X_α is a vertex of the graph $\Gamma_Y(A)$. But $i \in X_\alpha$. Thus we have $i \in \widetilde{Y}$ so that

$$\bigcup_k \operatorname{supp} A^k f \subseteq \widetilde{Y}.$$

Now we prove the converse inclusion. Let $i \in \widetilde{Y}$ and X_α be the vertex of the graph $\Gamma(A)$ that is determined by the index i. Since $i \in \widetilde{Y}$, we see that either $X_\alpha \cap Y \neq \varnothing$ or there exists a vertex X_β of the graph $\Gamma(A)$ such that $X_\beta \cap Y \neq \varnothing$ and there is a path in $\Gamma(A)$ starting from X_α and ending in X_β. Further assume that $X_\beta = X_\alpha$, when $X_\alpha \cap Y \neq \varnothing$. Let $j \in X_\beta \cap Y$. If $j = i$ then

$$i \in Y = \operatorname{supp} f \subseteq \bigcup_k \operatorname{supp} A^k f.$$

For $j \neq i$ it follows from the above that there exist a number $k \in N$ and indices $i_1, i_2, \dots, i_{k-1}$ such that $a_{ii_1} \odot a_{i_1 i_2} \odot \cdots \odot a_{i_{k-1} j} > \mathbf{0}$. Hence $a_{ij}^{(k)} > \mathbf{0}$. Since $Y = \operatorname{supp} f$ and $j \in Y$, we have $f_j > \mathbf{0}$ and so

$$(A^k f)_i = \bigoplus_{j=1}^{n} a_{ij}^{(k)} \odot f_i > \mathbf{0}.$$

Therefore, $i \in \operatorname{supp} A^k f$ and so $i \in \bigcup_k \operatorname{supp} A^k f$. The assertion is proved.

Assertion 7.4.1 implies that the semimodule

$$\mathscr{B}(\widetilde{Y}, K) = \{f \in K^n \mid f_i = \mathbf{0},\ \forall i \in X \setminus \widetilde{Y}\}$$

is invariant with respect to the endomorphism A. From the definition of the spectral function it follows that $\lambda_Y(i) = \mathbf{0}$ for all $i \in X \setminus \widetilde{Y}$, and therefore, Assertion 7.4.1 implies inequality (2) for all $i \in X \setminus \widetilde{Y}$. Hence, without loss of generality it can be assumed further that $\widetilde{Y} = X$; otherwise it is sufficient to restrict the endomorphism A to the subsemimodule $\mathscr{B}(\widetilde{Y}, K)$.

ASSERTION 7.4.2. *Let the endomorphism A acting in K^n be the direct sum of endomorphisms A_I, i.e. $A = \dot{+}_{I=1}^{k} A_I$, and $X = \bigcup X_I$ be the decomposition of the set $X = \{1, 2, \dots, n\}$ corresponding to A. For an arbitrary set $Y \subseteq X$ denote by λ_Y^I the spectral function of the endomorphism A_I on the set X_I. If λ_Y is the spectral function of the endomorphism A, then the following equalities hold*:

i) $\lambda_Y(x) = \bigoplus \lambda_Y^I(x)$;

ii) $\lambda_Y^I(x) \odot \lambda_Y^J(x) = \mathbf{0}$.

The definition of the spectral function immediately implies the proof.

Assertions 7.2 and 7.4.2 show that it is sufficient to prove Theorem 7.1 in the case of irreducible endomorphisms.

ASSERTION 7.4.3. *Suppose $A\colon K^n \to K^n$ is an arbitrary endomorphism acting in K^n. Then for every element $f \in K$ and $x \in X = \{1, 2, \dots, n\}$ we have $\lambda_Y(x) = \lambda_{Y^{(1)}}(x)$, where $Y = \operatorname{supp} f$, $Y^{(1)} = \operatorname{supp} Af$, $\lambda_Y(x)$ and $\lambda_{Y^{(1)}}(x)$ are the spectral functions of the endomorphism A corresponding to the sets Y and $Y^{(1)}$ respectively.*

PROOF. Assertion 7.4.1 yields the inequality $\lambda_Y(x) \geqslant \lambda_{Y^{(1)}}(x)$. Now we prove the inverse inequality: $\lambda_Y(x) \leqslant \lambda_{Y^{(1)}}(x)$. Let $x \in X_\alpha$ be an arbitrary point. If X_α is not a vertex of the graph $\Gamma_Y(A)$, then $\lambda_Y(x) = \lambda_{Y^{(1)}}(x) = \mathbf{0}$. So we shall assume that X_α is a vertex of $\Gamma_Y(A)$. We have

$$\lambda_Y(x) = \sup\{\lambda_\beta \mid X_\beta \in \Gamma_{X_\alpha, Y}(A)\},$$
$$\lambda_{Y^{(1)}}(x) = \sup\{\lambda_\beta \mid X_\beta \in \Gamma_{X_\alpha, Y^{(1)}}(A)\}.$$

These equalities imply that the inequality $\lambda_Y(x) \leqslant \lambda_{Y^{(1)}}(x)$ follows from the inclusion $X_\beta \in \Gamma_{X_\alpha, Y^{(1)}}(A)$ if only $X_\beta \in \Gamma_{X_\alpha, Y}(A)$ and $\lambda_\beta \neq \mathbf{0}$.

Let $X_\beta \in \Gamma_{X_\alpha, Y}(A)$. Then either $X_\beta \cap Y \neq \varnothing$ or there exists a vertex $X_\gamma \in \Gamma(A)$ such that $X_\gamma \cap Y \neq \varnothing$ and there is a path starting from X_β and ending in X_γ. In the first case we have the following relations:

$$X_\beta \cap Y^{(1)} = X_\beta \cap \operatorname{supp} Af \supseteq X_\beta \cap \operatorname{supp} A_\beta(f|_{X_\beta}) = \operatorname{supp} A_\beta(f|_{X_\beta}),$$

where A_β is the endomorphism from Definition 7.3. Since $X_\beta \cap Y \neq \varnothing$, we see that $f|_{X_\beta} \neq \mathbf{0}$; now from Assertion 7.2 and the condition $\lambda_\beta \neq \mathbf{0}$ we obtain $A_\beta(f|_{X_\beta}) \neq \mathbf{0}$. Therefore $X_\beta \cap Y^{(1)} \supseteq \operatorname{supp} A_\beta(f|_{X_\beta}) \neq \varnothing$ and this implies $X_\beta \in \Gamma_{X_\alpha, Y^{(1)}}(A)$.

Now consider the other case. Let $j \in X_\gamma \cap Y$. The existence of a path starting from X_β and ending in X_γ implies the existence for any $i \in X_\beta$ of a sequence of indices $i_1, i_2, \dots, i_s$ such that

$$a_{ii_1} \odot a_{i_1 i_2} \odot \cdots \odot a_{i_{s-1} i_s} \odot a_{i_s j} > \mathbf{0}. \tag{8}$$

Since $j \in X_\gamma \cap Y$, we have $f(j) > \mathbf{0}$, and thus, $a_{i_s j} \odot f(j) > \mathbf{0}$. Hence $(Af)(i_s) \geqslant a_{i_s j} \odot f(j) > \mathbf{0}$, i.e. $i_s \in \operatorname{supp} Af = Y^{(1)}$. Let $i_s \in X_\delta$. Then $i_s \in X_\delta \cap Y^{(1)}$. The last inclusion and (8) imply $X_\delta \cap Y^{(1)} \neq \varnothing$. Therefore $X_\beta \in \Gamma_{X_\alpha, Y^{(1)}}(A)$. Assertion 7.4.3 is proved.

ASSERTION 7.4.4. *Let $A\colon K^n \to K^n$ be an arbitrary irreducible endomorphism. Then for every element $f \in K^n$ such that* $\operatorname{supp} f = \operatorname{supp} Af = X$ *there exist an element $m_f \in K$ ($m_f \neq \mathbf{0}$) and a natural number $k_0 \in \mathbb{N}$ such that for all $k \geqslant k_0$ and $i \in X$ the inequalities*

$$(A^k f)(i) \geqslant m_f \odot \lambda_X^k(i) \tag{9}$$

hold, where $\lambda_X(i)$ is the spectral function of the endomorphism A.

PROOF. Let $X = \bigcup_{I=1}^m X_I$ be the decomposition of the set X and $\Gamma(A)$ be the graph corresponding to A. Renumber the vertices X_I in such a way that X_I and X_J are joined in $\Gamma(A)$ by an edge starting at X_I and ending at X_J only when $I \leqslant J$ (the possibility of such a renumbering follows from Assertion 7.2). To the decomposition of the set X there corresponds a direct sum decomposition of $K^n = \mathscr{B}(X, K)$ of the form

$$K^n = \dot{+}_{I=1}^{m} \mathscr{B}(X_I, K) = \dot{+}_{J=1}^{m} K^{|X_I|}.$$

Then every element $f \in K^n$ can be written in the form $f = (f_1, \dots, f_m)$, where $f_I \in \mathscr{B}(X_I, K)$. Now define the endomorphisms

$$A_{IJ}\colon \mathscr{B}(X_J, K) \to \mathscr{B}(X_I, K) \qquad (I \neq J)$$

by putting $A_{IJ} f = (AT_J f)|_{X_I}$, where $T_J\colon \mathscr{B}(X_J, K) \to \mathscr{B}(X, K)$ is the mapping given by

$$(T_J f)(x) = \begin{cases} f(x) & \text{if } x \in X_J, \\ \mathbf{0} & \text{if } x \notin X_J. \end{cases}$$

The endomorphisms A_{IJ} are zero for $I > J$. Thus the endomorphism A can be represented in the matrix form as follows:

$$A = \begin{bmatrix} A_1 & A_{12} & A_{13} & \dots & A_{1m} \\ \mathbf{0} & A_2 & A_{23} & \dots & A_{2m} \\ \mathbf{0} & \mathbf{0} & A_3 & \dots & A_{3m} \\ \vdots & \vdots & \vdots & \ddots & \vdots \\ \mathbf{0} & \mathbf{0} & \mathbf{0} & \dots & A_m \end{bmatrix},$$

where $A_J: \mathscr{B}(X_J, K) \to \mathscr{B}(X_J, K)$ are the endomorphisms from 7.3, and hence the endomorphism A^k can be represented in the matrix form as

$$A^k = \begin{bmatrix} A_1^{(k)} & A_{12}^{(k)} & A_{13}^{(k)} & \dots & A_{1m}^{(k)} \\ \mathbf{0} & A_2^{(k)} & A_{23}^{(k)} & \dots & A_{2m}^{(k)} \\ \mathbf{0} & \mathbf{0} & A_3^{(k)} & \dots & A_{3m}^{(k)} \\ \vdots & \vdots & \vdots & \ddots & \vdots \\ \mathbf{0} & \mathbf{0} & \mathbf{0} & \dots & A_m^{(k)} \end{bmatrix},$$

where

$$A_I^{(k)} = A_I^k, \tag{10}$$
$$A_{IJ}^{(k)} = {\bigoplus}' A_I^{\alpha_0} \circ A_{II_1} \circ A_{I_1}^{\alpha_1} \circ A_{I_1 I_2} \circ \dots \circ A_{I_l J} \circ A_J^{\alpha_{l+1}} \bigoplus_{\substack{p+q=k-1 \\ p,q \geqslant 0}} A_I^p \circ A_{ij} \circ A_J^q;$$

here ${\bigoplus}'$ is the summation over the set of indices $I_1, \dots, I_l$ such that

$$I_1 > I_2 > \dots > I_l, \quad \sum_{i=0}^{l+1} \alpha_i + l + 1 = k, \quad \alpha_i \geqslant 0.$$

Now we prove (9). Suppose $i \in X_I$ is an arbitrary index. Then

$$(A^k f)(i) = \bigoplus_{I=J}^{m} (A_{IJ}^{(k)} f_J)(i) \geqslant (A_{IJ}^{(k)} f_J)(i)$$

for all indices J such that $X_J \in \Gamma_{X_I, X}(A)$. Let us fix an index J. From Proposition 7.1 and the condition $\operatorname{supp} f_J = X_J$ we obtain the inequality $(A_{IJ}^{(k_0)} f_J)(i) > \mathbf{0}$ which is true for some $k_0 \in \mathbb{N}$. Then for all $s > 0$ the estimate

$$(A^{k_0+s} f)(i) = (A^{k_0} \circ A^s f)(i) \geqslant (A_{IJ}^{(k_0)} \circ A_J^s f_J)(i)$$

is satisfied. From inequality (6) we get

$$(A_{IJ}^{(k_0)} \circ A_J^s f_J)(i) \geqslant m_J \odot \lambda_J^{k_0+s} \quad \text{for all } s > 0,$$

where

$$m_J = m_{f_J} \odot (A_{IJ}^{(k_0)} f_j)(i) \odot \lambda_J^{-k_0} \neq \mathbf{0}.$$

(Here and everywhere below we assume that $\lambda_J \neq \mathbf{0}$, because whenever $\lambda_J = \mathbf{0}$ the proof is obvious.) Thus for all indices J considered and k sufficiently large we have the inequalities $(A^k f)(i) \geqslant m_J \odot \lambda_J^k$ which imply the inequality

$$(A^k f)(i) \geqslant \bigoplus_{X_J \in \Gamma_{X_I, X}(A)} m_J \odot \lambda_J^k.$$

For k sufficiently large the last inequality and the stabilization condition yield estimate (9). The proof is complete.

ASSERTION 7.4.5. *Let $A: K^n \to K^n$ be an arbitrary irreducible endomorphism and λ be its eigenelement. Then there exists an element $M \in K$ such that for all $f \in K^n$ and $k = 0, 1, \dots$ we have the inequalities*

$$\|A^k f\| \leqslant M \odot \lambda^k \odot \|f\|,$$

where $\|f\| = \sup_i f(i)$.

The proof is analogous to that of Assertion 7.1.

ASSERTION 7.4.6. *Let $A: K^{n_1} \to K^{n_2}$ be an arbitrary homomorphism defined by the matrix (a_{ij}). Suppose $\|A\| = \sup_{i,j} a_{ij}$. Then for all $f \in K^{n_1}$ we have the inequality $\|Af\| \leqslant \|A\| \odot \|f\|$.*

The proof is evident.

ASSERTION 7.4.7. *Let $A: K^n \to K^n$ be an arbitrary irreducible endomorphism. Then for any element $f \in K^n$ such that $\operatorname{supp} f = \operatorname{supp} Af = X$ there exist an element $M_f \in K$ and a natural number $k_0 \in \mathbb{N}$ such that for all $k \geqslant k_0$ and $i \in X$ we have the inequalities*

$$(A^k f)(i) \leqslant M_f \odot \lambda_X^k(i), \tag{11}$$

where $\lambda_X(i)$ is the spectral function of the endomorphism A.

PROOF. It is clear that we can assume that $\|A\| \leqslant \mathbf{1}$. If $i \in X_I$ is an arbitrary index, then

$$(A^K f)(i) = \bigoplus_{I=J}^{m} (A_{IJ}^{(k)} f_J)(i). \tag{12}$$

From Proposition 7.1 it follows that if a vertex X_J does not belong to the graph $\Gamma_{X_I,X}(A)$, then $A_{IJ}^{(k)} = \mathbf{0}$. Therefore in (12) the summation is over all indices such that $X_J \in \Gamma_{X_I,X}(A)$. Now we give an estimate of every term on the right-hand side part of equation (12). First, we have the inequality $(A_{IJ}^{(k)} f_J)(i) \leqslant \|A_{IJ}^{(k)} f_J\|$. Next, equality (10), Assertions 7.4.5, 7.4.6, Definition 7.3 and the inequality $\|A_{IJ}^{(k)}\| \leqslant \mathbf{1}$ imply the following:

$$\begin{aligned}
\|A_{IJ}^{(k)} f_J\| &= {\bigoplus}' \|A_I^{\alpha_0} \circ A_{II_1} \circ A_{I_1}^{\alpha_1} \circ A_{I_1 I_2} \circ \cdots \circ A_{I_l J} \circ A_J^{\alpha_{l+1}} f_J\| \\
&\quad \times \bigoplus_{\substack{p+q=k-1 \\ p,q \geqslant 0}} \|A_J^p \circ A_{IJ} \circ A_J^q f_J\| \\
&\leqslant M_1 \odot \left[{\bigoplus}' (\lambda_I^{\alpha_0} \odot \lambda_{I_1}^{\alpha_1} \odot \cdots \odot \lambda_J^{\alpha_{l+1}} \odot \|f_J\|) \bigoplus_{\substack{p+q=k-1 \\ p,q \geqslant 0}} \lambda_I^p \odot \lambda_J^q \odot \|f_J\| \right] \\
&\leqslant M_1 \odot \left[\bigoplus (\lambda_X(i))^{\alpha_0+\alpha_1+\cdots+\alpha_{l+1}} \bigoplus (\lambda_X(i))^{k-1} \right] \odot \|f_J\| \\
&\leqslant M \odot \lambda_X^k(i),
\end{aligned}$$

where $M_1, M \in K$ are constants and $\bigoplus'$ is the summation over the set of indices $I_1, \dots, I_l$ satisfying the following conditions:

$$I_1 > I_2 > \cdots > I_l \quad \text{and} \quad \sum_{i=0}^{l+1} \alpha_i + l + 1 = k, \ \alpha_i \geqslant 0.$$

Finally, the last inequalities and equality (12) immediately imply the required estimate.

REMARK 7.1. In Theorem 7.1 the condition $\operatorname{supp}(A^{s-1} f) = \operatorname{supp}(A^s f)$ can be replaced by one of the following conditions:

i) $\operatorname{supp} f = X$;

ii) $\operatorname{supp} f = \bigcup_{\alpha \in I} X_I$ and $\lambda_\alpha \neq \mathbf{0}$, $\forall \alpha \in I$ (I is some subset of the set $\{1, \dots, m\}$).

THEOREM 7.2. *Suppose $A\colon K^n \to K^n$ is an arbitrary irreducible endomorphism and for some element $f \in K^n$ there exist elements $m, M \in K$ $(m \neq \mathbf{0})$, a function $\alpha\colon X \to K$, and a natural number k_0 such that for all $k \geqslant k_0$ and $i \in X$ the estimate*

$$m \odot \alpha^k(i) \leqslant (A^k f)(i) \leqslant M \odot \alpha^k(i) \tag{13}$$

holds. Then:

i) *there exists a number $s \in \mathbb{N}$ such that* $\operatorname{supp}(A^{s-1} f) = \operatorname{supp}(A^s f)$;

ii) *$\alpha(\cdot)$ is the spectral function $\lambda_Y(\cdot)$ of the endomorphism A, where* $Y = \operatorname{supp} f$.

PROOF. Assertion i) immediately follows from (13). Let us prove ii). Suppose $\lambda_Y(\cdot)$ is the spectral function of the endomorphism A corresponding to the set $Y = \operatorname{supp} f$. Assertion i) and Theorem 7.1 imply that there exist elements $m_f, M_f \in K$ $(m \neq \mathbf{0})$ such that for k sufficiently large estimate (7) is satisfied. Then from (7) and (13) for k sufficiently large we obtain

$$m \odot \alpha^k(i) \leqslant M_f \odot \lambda_Y^k(i), \qquad m_f \odot \lambda_Y^k(i) \leqslant M \odot \alpha^k(i).$$

From the last inequalities and the stabilization condition it follows that $\alpha(i) = \lambda_Y(i)$. Hence the proof is over.

REFERENCES

1. V. P. Maslov, *Asymptotic methods for solving pseudodifferential equations*, "Nauka", Moscow, 1987. (Russian)

2. S. M. Avdoshin, V. V. Belov, and V. P. Maslov, *Mathematical aspects of the design of computational media*, Moscow Institute of Electronic Engineering, Moscow, 1984. (Russian)

3. V. N. Kolokoltsov and V. P. Maslov, *Idempotent calculus as the apparatus of optimization theory*, I, Funktsional. Anal. i Prilozhen. **23** (1989), no. 1, 1–14; II, **23**, no. 4, 53–62; English transl. in Functional Anal. Appl. **23** (1989).

4. P. I. Dudnikov and S. N. Samborskiĭ, *Endomorphisms of semimodules over semirings with an idempotent operation*, Preprint No. 87-48, Inst. Mat. Akad. Nauk Ukrain., Kiev, 1987. (Russian)

5. ______, *Endomorphisms of semimodules over semirings with an idempotent operation*, Izv. Akad. Nauk SSSR Ser. Mat. **55** (1991), 93–109; English transl. in Math. USSR-Izv. **38** (1992).

6. I. V. Romanovskiĭ, *Optimization of stationary control of discrete deterministic process in dynamic programming*, Kibernetika (Kiev) **2** (1967), 66–78; English transl. in Cybernetics **3** (1967).

7. ______, *Asymptotic behavior of a discrete deterministic process with continuous state space*, Optimal. Planirovanie **8** (1967), "Nauka", Novosibirsk, 171–193. (Russian)

CENTER FOR OPTIMIZATION AND MATHEMATICAL MODELING, INSTITUTE OF NEW TECHNOLOGIES, KIROVOGRADSKAYA UL., 11, 113587, MOSCOW, RUSSIA

Translated by I. PETRENKO

ADVANCES IN SOVIET MATHEMATICS
Volume 13, 1992

On Linear, Additive, and Homogeneous Operators in Idempotent Analysis

V. N. KOLOKOLTSOV

§1. Introduction

This whole volume is devoted to the investigation of linear operators (or endomorphisms) and of linear equations on the space of continuous functions with values in a semiring with idempotent addition. In §1 of this article we first study compact and invertible operators. We obtain a characterization of these classes of operators in terms of their integral kernel. We prove, in particular, that the group of invertible operators in idempotent calculus is rather poor as compared to the case of ordinary linear operators. A general invertible operator proves to be the composition of a multiplication operator by a function and of an operator generated by a homeomorphism of the base (this theorem was announced in [1, 2]). As a consequence, we see that all automorphisms of the semialgebra of linear operators are inner. Note that the group of invertible operators may be exploited to transform an operator to a simpler but equivalent form (see, for example, [3]). In §1 we also prove a version of a theorem on the existence of an eigenvalue and eigenfunction for the general compact operator and use these results for an investigation of a certain stationary Hamilton-Jacobi equation.

The interest in the investigation of idempotent linear equations is connected with the fact that such equations describe a wide class of deterministic problems in optimal control theory. Yet not all optimization problems are included in this class. Therefore, further in this article we study two more general classes of operators: additive and homogeneous. We restrict ourselves to the most important case of the numerical semiring with operations $\oplus = \min$ and $\odot = +$. Thus an operator B on the set of numerical functions will be called *additive* (respectively, *homogeneous*), if

$$B(\min(h_1, h_2)) = \min(B(h_1), B(h_2))$$

for all functions h_1, h_2 (respectively, $B(a+h) = a + B(h)$ for all numbers a and functions h).

1991 *Mathematics Subject Classification*. Primary 16Y60, 47B38; Secondary 90C27.

Let us first present two examples where additive operators naturally appear. In discrete optimization theory the additive Bellman operator describes trajectory type problems with variable time of motion. Namely, let $G = (X, E)$ be a graph with the set of vertices X and the set of arcs E. Let $\varphi_e(t)$ define the time of motion along the arc $e \in E$ if this motion begins at the moment of time t. The problem considered is to construct the shortest path from the fixed point $x_0 \in X$ to any other point $x \in X$, if the motion begins at time t_0. The corresponding Bellman operator B acts on the set of functions $h\colon X \to \{t \in \mathbb{R}, t \geqslant t_0\}$ by the formula

$$Bh(x_i) = \min_e \min_{\tau \geqslant h(x_i)} \{\tau + \varphi_e(\tau)\},$$

where the first minimum is taken among all arcs $e \in E$ entering x_i; the shortest time $S(x)$ of motion from x_0 to x satisfies the following additive (in the sense of the operation $\oplus = \min$) Bellman equation

$$S = BS \oplus F,$$

where F is the characteristic function of the point x_0, i.e., F is equal t_0 at x_0 and to $+\infty$ at other points $x \in X$.

Another important example of an additive operator is the solving operator of the Cauchy problem for first order partial differential equations. Namely, let the function $H(x, p, S)$ be convex with respect to p and $L(x, v, S)$ be its Legendre transform with respect to this variable. Under rather general assumptions, the solution $S(t, x)$ of the Cauchy problem for the equation

$$\frac{\partial S}{\partial t} + H\left(x, \frac{\partial S}{\partial x}, S\right) = 0 \tag{1.1}$$

with the initial function $S_0(x)$, $x \in \mathbb{R}^n$, is given by the family of the operators R_t of the form

$$S(t, x) = R_t(x) = \min_{\xi \in \mathbb{R}^n} G(t, x, \xi, S_0(\xi)),$$

where $G(t, x, \xi, a)$ is the infimum of the functional $S(t, a, x(\cdot))$ among all continuous piecewise smooth curves $x(\tau)$ with endpoints $x(0) = \xi$, $x(\tau) = x$. Here $S(t, a, x(\cdot))$ denotes the value $S(t)$ of the solution of the ordinary differential equation $\dot{S} = L(x, \dot{x}, S)$ with the initial data $S(0) = a$. It is evident that the operator R_t is additive with respect to the operation $\oplus = \min$. The existence of this superposition principle allows one to define the generalized solutions of the above Cauchy problem in a natural way, extending the definition from the classical solutions by continuity and additivity (see §2). If the Hamiltonian $H(x, p, S)$ does not depend on S, then equation (1.1) is the Hamilton-Jacobi equation. In that case the solving operator R_t becomes linear with respect to the operations $\oplus = \min$ and $\odot = +$ [4–6]. In [19] an analog of the Bellman equation is derived which describes the dynamics of Pareto set in multicriterial optimization problems.

Let us note that in ordinary arithmetic each additive continuous function is linear. Evidently, this is not so in idempotent arithmetic, because each monotone function $h: \mathbb{R} \mapsto \mathbb{R}$ satisfies the condition

$$h(\min(a, b)) = \min(h(a), h(b)),$$

but it need not be linear with respect to the operations $\oplus = \min$ and $\odot = +$. Yet in some sense one can reduce the solution of an additive problem in idempotent calculus to the solution of a linear one. In the case of the space of functions defined on a finite state space this reduction was used in [7, 8] for constructing algorithms that solve the trajectory problem with variable time of motion described above. The following mathematical fact is the basis of this reduction. Let M be an idempotent semigroup, $\mathrm{End}(M)$ be the semiring of its endomorphisms with composition as the multiplication, and X be a compact set. Then the general representation theorem of additive operators allows us to show (see §2) that the semiring of endomorphisms of the semigroup $C(X, M)$ of continuous functions $X \to M$ is isomorphic to the semiring of endomorphisms of the semimodule $C(X, \mathrm{End}(M))$.

The appearance of the homogeneous Bellman operator in optimization theory is usually related to conflict control and to stochastic control. When investigating these operators, we restrict ourselves to the case of a finite state space for simplicity. In §2 we shall show that the set of all maps $B: \mathbb{R}^n \to \mathbb{R}^n$ satisfying the conditions

(i) $B(a+h) = a + B(h)$ for all $a \in \mathbb{R}$, $h \in \mathbb{R}^n$, and
(ii) $\|B(h) - B(g)\| \leqslant \|h - g\|$ for all $h, g \in \mathbb{R}^n$ (here $\|h\| = \max |h^i|$ for $h \in \mathbb{R}^n$)

coincide with the set of all Bellman operators for different multistep antagonistic games on the state space $X = \{1, \dots, n\}$. If B also satisfies the additional convexity property

$$B(a_1 h_1 + a_2 h_2) \leqslant a_1 B(h_1) + a_2 B(h_2), \qquad a_1, a_2 \geqslant 0, \ a_1 + a_2 = 1,$$

where the sign $\leqslant$ denotes the Pareto order, then B is the Bellman-Howard operator of some Markov decision process on the state space X. A similar fact can be proved for more general state spaces.

For a homogeneous operator B, as well as for a linear one, it is natural to consider the equation $B(h) = a + h$, which is the analog of the equation for eigenvectors from usual linear theory. It turns out that the solutions of this equation define the mean income and the optimal stationary strategies for long time optimal processes and sometimes the turnpikes as well. For Markov decision processes, these results were obtained, for example, in [9–11]. In §3 of the article, we get similar results for general antagonistic multistep games. In [18] perturbation theory is developed; it gives approximate solutions of the equation $B(h) = a + h$ for a stochastic Bellman equation.

§2. General properties of idempotent linear operators. Invertible and compact operators. The spectrum

Let $A = \mathbb{R} \cup \{+\infty\}$ be the semiring with operations $\oplus = \min$, $\odot = +$, neutral elements $\mathbf{0} = +\infty$, $\mathbf{1} = 0$ and the metric $\rho(a, b) = |e^{-a} - e^{-b}|$. Let X be a locally compact set and $C_{\mathbf{0}}(X)$ be the space (or, more exactly, semimodule) of continuous functions $f\colon X \to A$ tending to $\mathbf{0}$ as $x \to \infty$ (i.e., $\rho(f(x), \mathbf{0}) < \varepsilon$ for any fixed ε and for all x outside some compact set $K \subset X$). The topology in $C_{\mathbf{0}}(X)$ is defined by the uniform metric. It was proved in [12, 6] that any linear operator $B\colon C_{\mathbf{0}}(X) \to C_{\mathbf{0}}(X)$ (i.e., any continuous map B, satisfying the condition

$$B(a_1 \odot h_1 \oplus a_2 \odot h_2) = a_1 \odot B(h_1) \oplus a_2 \odot B(h_2)$$

for all $a_1, a_2 \in A$, $h_1, h_2 \in C_{\mathbf{0}}(X)$) can be represented in the form

$$Bh(x) = \inf_y (b(x, y) \odot h(y)),$$

where $b\colon X \times X \to A$ is a semicontinuous function bounded from below in both arguments such that $b(x, y) \to \mathbf{0}$ as $x \to \infty$ uniformly in y from any compact set. The function b is called the *integral kernel* of the operator B.

First of all we give the description of two important classes of linear operators, namely invertible and compact operators.

THEOREM 1. *Let $B\colon C_{\mathbf{0}}(Y) \to C_{\mathbf{0}}(X)$ and $D\colon C_{\mathbf{0}}(X) \to C_{\mathbf{0}}(Y)$ be mutually inverse operators (X, Y are locally compact sets). Then there exist a homeomorphism $\beta\colon X \to X$ and continuous functions $\varphi\colon X \to A$, $\psi\colon Y \to A$ such that*

$$\varphi(x) \odot \psi(\beta(x)) = \mathbf{1} \tag{2.1}$$

for all $x \in X$ and B, D are defined by formulas

$$Bh(x) = \varphi(x) \odot h(\beta(x)), \tag{2.2}$$

$$Dg(y) = \psi(y) \odot g(\beta^{-1}(y)). \tag{2.3}$$

PROOF. Writing out the assumptions of the theorem in terms of the integral kernels b, d of the operators B, D we get

$$\begin{aligned} \inf_y (b(x, y) \odot d(y, z)) &= \begin{cases} \mathbf{1}, & x = z, \\ \mathbf{0}, & x \neq z, \end{cases} \\ \inf_x (d(y, x) \odot b(x, t)) &= \begin{cases} \mathbf{1}, & y = t, \\ \mathbf{0}, & y \neq t. \end{cases} \end{aligned} \tag{2.4}$$

It follows immediately that for each $x \in X$ there exists a $y(x) \in Y$ such that $b(x, y(x)) \neq \mathbf{0}$, $d(y(x), x) \neq \mathbf{0}$, and for each $y \in Y$ there exists an $x(y) \in X$ such that $d(y, x(y)) \neq \mathbf{0}$, $b(x(y), y) \neq \mathbf{0}$. Moreover, these $x(y)$ and $y(x)$ are defined uniquely, for if there exist $y_1 \neq y_2$ such that $d(y_i, x) \neq \mathbf{0}$, then

$$\inf_x (d(y_i, x) \odot b(x, y(x))) \neq \mathbf{0},$$

which contradicts (2.4). Thus, there exists a bijective map $\beta: X \to Y$ such that

$$b(x, y) \neq \mathbf{0} \iff d(y, x) \neq \mathbf{0} \iff y = \beta(x).$$

Now it follows from (2.4) that the functions $\varphi(x) = b(x, \beta(x))$ and $\psi(y) = d(y, \beta^{-1}(y))$ satisfy (2.1) and equations (2.2) and (2.3) hold.

COROLLARY 1. *If the topological modules $C_0(X)$ and $C_0(Y)$ are isomorphic, then the topological sets X and Y are homeomorphic.*

REMARK. As is well known, the same result holds also for ordinary spaces of continuous real functions. What is rather surprising is that the assertion of Corollary 1 remains true if we consider the space of continuous A-valued functions on X in the weak topology (the topology of pointwise convergence) [2]. A similar statement does not hold for the space of ordinary continuous real-valued functions with weak topology.

It follows also from Theorem 1 that each automorphism of the operator semialgebra is inner.

COROLLARY 2. *Each automorphism F (i.e., any continuous and bijective map preserving idempotent addition $\oplus$ and composition) of the semialgebra of linear operators in $C_0(X)$ takes an operator B with integral kernel b to the operator $F(B)$ with integral kernel $\bar{b}$ defined by the formula*

$$\bar{b}(x, y) = \varphi(x) - \varphi(y) + b(\beta(x), \beta(y)),$$

where β is a homeomorphism of X and $\varphi: X \to A \setminus \{\mathbf{0}\}$ is a continuous function, i.e., $F(B) = C \circ B \circ C^{-1}$ for an invertible C.

We omit the simple proof based on the application of Theorem 1.

Another important class of operators is the class of compact operators. The linear operator $B: C_0(Y) \to C_0(X)$ is called *compact* if it takes each bounded (in the sense of the metric) set of continuous functions into a precompact one. Let us suppose that the locally compact set X is separable, and therefore, its topology can be determined by the metric d on it. A simple application of the Arzela theorem gives the following result.

THEOREM 2. *The linear operator B is compact if and only if its integral kernel $b(x, y)$ is uniformly continuous with respect to the first argument (i.e., for all $\varepsilon > 0$ there exists a $\delta > 0$ such that*

$$d(x_1, x_2) < \delta \implies \rho(b(x_1, y), b(x_2, y)) < \varepsilon$$

for all $y \in Y$) and $b(x, y) \to \mathbf{0}$ as $x \to \infty$ uniformly in y.

COROLLARY 3. *The compact operator $B: C_0(Y) \to C_0(X)$ can be extended to the space of all bounded A-valued functions on Y, and B maps this space to $C_0(X)$.*

Now we are going to prove a theorem on the existence of the spectrum for compact operators. Similar results in implicit form or under more restrictive assumptions were obtained, for example, in [13, 9]. The existence of the spectrum for indecomposable operators with continuous integral kernel in the case of a compact set X and more general idempotent semirings (with the contraction law) was proved in [14] by nonstandard calculus methods.

Let B be a compact operator in $C_{\mathbf{0}}(X)$ with integral kernel b. Let d be a metric on the locally compact set X and μ be a probability measure on X such that the measure of any open set is positive. We denote $D = \inf b(x, y)$. It follows from Theorem 2 that $D > -\infty$ (i.e., $D \in A$).

To state the theorem, we need additional nondegeneracy properties:

A) $\forall y \, \exists x : \; b(x, y) \neq \mathbf{0}$,
B) $\exists C \neq \mathbf{0} : \; \forall y, \; \inf_x b(x, y) \leqslant C$,
C) the set $\{y \in X \mid \exists x : b(x, y) = \mathbf{0}\}$ has no interior points.

Evidently, A) $\Rightarrow$ B), if X is a compact set and the integral kernel b is continuous.

THEOREM 3. *Let B satisfy condition* B). *Then there exist $a \neq \mathbf{0}$ and $h \in C_{\mathbf{0}}(X)$, h not being identically $\mathbf{0}$, such that*

$$B(h) = a \odot h = a + h.$$

If B also satisfies C), *then the eigenvalue a is unique and the eigenfunction h does not vanish anywhere.*

PROOF. The uniqueness in the case of C) follows from standard arguments that use the adjoint operator. So we restrict ourselves to the proof of the existence theorem under condition B).

Let us consider the map $E \colon h(x) \mapsto Eh(x) = e^{-h(x)}$, which is an isometry from $C_{\mathbf{0}}$ to the cone $C_+(X)$ of nonnegative functions in the Banach space $C(X)$ of real continuous functions on X vanishing at infinity. The conditions of the theorem imply that the continuous map

$$M = E \circ B \circ E^{-1} \colon C_+(X) \to C_+(X)$$

has the following properties:

a) M is positively homogeneous: $\forall \lambda > 0$, $g \in C_+(X)$, $M(\lambda g) = \lambda M(g)$;
b) M satisfies the estimates: $e^{-C}\|g\| \leqslant \|M(g)\| \leqslant e^{-D}\|g\|$ for all $g \in C_+(X)$, which implies, in particular, that $M^{-1}(0) = 0$,
c) for all $\varepsilon > 0$ there exists a $\delta > 0$ such that

$$d(x_1, x_2) < \delta \implies \forall g, \; |Mg(x_1) - Mg(x_2)| \leqslant \varepsilon\|g\|.$$

Let us introduce an equivalence relation on $C_+(X)\backslash\{0\}$: $v \sim g$, if $v = \lambda g$ for some $\lambda > 0$. Let Φ be the corresponding quotient set and $\widetilde{M} \colon \Phi \to \Phi$ be the quotient map taking the ray $\{\lambda g, \lambda > 0\}$ to the ray $\{\lambda M(g), \lambda > 0\}$. It is clear that Φ can be identified with the part of the unit sphere $\{g : \|g\| = 1\}$

in $C_+(X)$ and the map $\widetilde{M}$ in this representation is defined by the formula

$$\widetilde{M} = \mathrm{pr} \circ M, \quad \text{where } \mathrm{pr}(h) = h/\|h\|.$$

Evidently, the existence of the eigenfunction and the eigenvalue of B is equivalent to the existence of a fixed point of the map $\widetilde{M}$. It follows from the two-sided estimates in b) that the image of $\widetilde{M}$ in Φ is a precompact set. But Φ is not a convex subset of $C_+(X)$ and an immediate application of the Leray-Schauder fixed point theorem is not possible. Thus, to prove the existence of the eigenvalue and of the eigenfunction of B it suffices to show that Φ is homeomorphic to a closed bounded convex subset of $C_+(X)$. For this purpose we use the probability measure μ on X defined above.

Let L be the hyperplane in $C(X)$, consisting of all h such that

$$\int_X h(x)\, d\mu(x) = 0.$$

We define the map $\Pi\colon \Phi \to L$ as the shift $\Pi(h) = h - \lambda(h)$, where $\lambda(h)$ is a constant (evidently, unique) such that $h - \lambda(h) \in L$. Since the measure μ of each open set in X is positive, it follows that $\lambda(h) \in (0, 1)$ for all $h \in \Phi$ and the map Π is continuous and injective. It is clear that the image

$$\Pi(\Phi) = \left\{ g \in L : \max_x g(x) \leqslant 1 + \min_x g(x) \right\}$$

is a closed subset of the unit ball $C(X)$. Let us show that $\Pi(\Phi)$ is convex. Let $g_1, g_2 \in \Pi(\Phi)$, and therefore

$$\max g_1 \leqslant 1 + \min g_1, \qquad \max g_2 \leqslant 1 + \min g_2.$$

Adding these two inequalities with coefficients α_1, α_2 such that $\alpha_j \geqslant 0$ and $\alpha_1 + \alpha_2 = 1$ we get

$$\alpha_1 \max g_1 + \alpha_2 \max g_2 \leqslant 1 + \alpha_1 \min g_1 + \alpha_2 \min g_2,$$
$$\max(\alpha_1 g_1 + \alpha_2 g_2) \leqslant 1 + \min(\alpha_1 g_1 + \alpha_2 g_2).$$

Thus any convex linear combination of two elements of $\Pi(\Phi)$ belongs to $\Pi(\Phi)$. Consequently, $\Pi(\Phi)$ is convex. Now we can use the Leray-Schauder theorem to get the existence of a fixed point for the map $\Pi \circ \widetilde{M} \circ \Phi^{-1}$ and, therefore, also for the map $\widetilde{M}\colon \Phi \to \Phi$. As we already pointed out, this concludes the proof of the theorem.

COROLLARY 1. *Let the operator B satisfy* B) *and* C) *so that the eigenvalue a is unique and the eigenfunction h never equals* $\mathbf{0}$. *Suppose $f \leqslant h + C$ for some constant C. Then*

$$\lim_{m\to\infty} B^m f(x)/m = a. \tag{2.5}$$

PROOF. Evidently, we have

$$\inf_x (f - h) \leqslant Bf(x) - Bh(x) \leqslant \sup_x (f - h)$$

for all x. Thus we get recursively

$$\inf_x(f-h) \leqslant B^m f(x) - B^m h(x) \leqslant \sup_x(f-h). \tag{2.6}$$

Hence we get (2.5), since $B^m(h) = ma + h$.

Note that this result is a slight generalization of that of [13].

Another important consequence of Theorem 3 is a sufficient condition for the convergence of the Neumann series. This series gives the so-called Duhamel solution [7, 8] of the equation $g = Bg \oplus f$ for the unknown function g, where f is a fixed function from $C_{\mathbf{0}}$.

COROLLARY 2. *Let the assumptions of Corollary 1 hold. Suppose $a > 0$ and $f \leqslant h + C$ for some constant C. Then the Neumann series*

$$f \oplus B(f) \oplus B^2(f) \oplus \cdots \tag{2.7}$$

is finite, i.e., it is equal to a finite sum $f \oplus B(f) \oplus \cdots \oplus B^k(f)$ for some k.

PROOF. It follows from (2.6) that

$$B^m f(x) - f(x) \geqslant \inf_x(f-h) + h(x) - f(x) + ma.$$

Hence

$$B^m f(x) \geqslant f(x) + \inf_x(f-h) + ma - C.$$

Therefore $B^m f(x)$ becomes greater than $f(x)$ for all x when m is sufficiently large. That implies the required statement.

REMARK. Note that the Neumann series (2.7) giving the Duhamel solution (which is evidently the maximal one) of the equation

$$g = Bg \oplus f \tag{2.8}$$

is widely used in discrete optimization problems.

As an application of the results obtained, we present a theorem on the solutions of the stationary Hamilton-Jacobi-Bellman equation. Let $H(x, p)$ be a smooth function on $\mathbb{R}^{2n}$ that is convex in p. Then the generalized solution $S(t, x) = (R_t S_0)(x)$ of the Cauchy problem

$$\frac{\partial S}{\partial t} + H\left(x, \frac{\partial S}{\partial x}\right) = 0, \qquad S|_{t=0} = S_0(x) \tag{2.9}$$

is given by the well-known formula (see, for instance, [6])

$$(R_t S_0)(x) = \inf_\xi(S_0(x) + S(t, x, \xi)),$$

where $S(t, x, \xi)$ is the infimum of the functional $J = \int_0^t L(x, \dot{x})\, d\tau$ defined on the set of continuous piecewise smooth curves joining ξ to x by the time t. Note that if $S(x) + t\lambda$ is a classical (i.e., smooth everywhere) solution of (2.9), then $S(x)$ satisfies the stationary equation

$$H(x, \partial S/\partial x) = \lambda. \tag{2.10}$$

Thus it is natural to define a *generalized solution* of (2.10) as a function $S(x)$ such that $(R_t S)(x) = S(x) + t\lambda$, i.e., as an eigenfunction of the operator R_t in the sense of the operations $\oplus = \min$, $\odot = +$. If the Lagrangian (which is the Legendre transformation in p of H) $L(x, \dot{x})$ is bounded from below and tends to infinity as $x, \dot{x} \to \infty$, then the operator R_t is compact (by Theorem 2) and there may exist only a unique λ for which (2.10) has a solution (by Theorem 3). The set of solutions of equation (2.10) with a fixed λ evidently forms a semimodule in the sense of operation $\oplus\odot$. Let us give an important example (the connection with exponential asymptotics is described in [20]) when this semimodule is a finite-dimensional one. Let us define the bilinear scalar product [4]

$$\langle f, h \rangle = \inf_x (f(x) + h(x)) \tag{2.11}$$

on the set of functions bounded from below.

THEOREM 3. *Suppose the Lagrangian $L(x, v)$ has only a finite number of minimum points $(\xi_1, 0), \ldots, (\xi_l, 0)$, say $L(\xi_j, 0) = 0$, $j = 1, \ldots, l$. Then the set of functions $S_j(x) = \inf_t S(t, x, \xi_j)$, $j = 1, \ldots, l$, forms a basis for the semimodule of the eigenfunctions of the operators R_t, and the set of functions $\widetilde{S}_j(x) = \inf_t S(t, \xi_j, x)$, $j = 1, \ldots, l$, forms a basis for the semimodule of the eigenfunctions of the adjoint operator $\widetilde{R}_t$ (in the sense of the product* (2.11)). *Moreover, for any real-valued function h bounded from below there exists a limit*

$$\lim_{t\to\infty} (R_t h)(x) = \bigoplus_{j,k} (h, \widetilde{S}_j) \odot (\widetilde{S}_j, S_k) \odot S_k(x). \tag{2.12}$$

The proof is based on the remark that the finite set $\xi_1, \ldots, \xi_l$ is an attractor for the extremal curves defined by the Lagrangian L.

§3. Additive operators and first order partial differential equations

The representation theorem for additive functionals [12] immediately implies the following result:

THEOREM 4. *Each additive operator B in C_0 (i.e., any continuous operator B such that $B(h \oplus g) = B(h) \oplus B(g)$ for all h, g) is defined by the formula*

$$Bh(x) = \inf_y b(x, y, h(y)) \tag{3.1}$$

for some lower semicontinuous function $b\colon X \times X \times A \to A$ that is monotone with respect to the last argument.

As a consequence we get

THEOREM 5. *There is an isomorphism between the semiring of additive operators in C_0 (with composition as the multiplication) and the semiring of linear operators in the space $C_0(X, \mathrm{End})$ of continuous functions $X \to \mathrm{End}$*

that vanish to $\mathbf{0}$ *at infinity, where* End *is the semiring of endomorphisms of the semigroup* $\mathbb{R} \cup \infty$ *with* $\oplus = \min$. *This isomorphism maps the operator* B *of the form* (3.1) *to the operator* $\widetilde{B}$ *such that*

$$(\widetilde{B}h)(x, m) = \inf_y b(x, y, h(y, m)) \quad \text{for } h \in C_0(X, \text{End}).$$

This correspondence was used in [8] in the case of finite X to get the convergence conditions for the Neumann series (2.7) with an arbitrary additive operator B. Using this correspondence, we shall now obtain the description of invertible additive operators. Namely, the proof of Theorem 1 remains valid, if we take the semiring End instead of the numerical semiring A. Thus we immediately get the following result:

THEOREM 6. *Each invertible additive operator in* C_0 *is defined by the formula*

$$Bh(x) = \varphi(x, h(x)),$$

for each x *where* φ *is an element of* End.

Now let us show how additive operators appear in the theory of first order partial differential equations.

Let $H(x, p, S)$ be a smooth function on $\mathbb{R}^n \times \mathbb{R}^n \times \mathbb{R}$ that is convex in p and let $L(x, v, S)$ denote its Legendre transform with respect to v. Let $G(t, x, \xi, a)$ be the infimum of the functional $S(t, a, x(\cdot))$ among all continuous piecewise smooth functions $x(\tau)$ with endpoints $x(0) = \xi$, $x(t) = x$, where $S(t, a, x(\cdot))$ denotes the value $S(t)$ of the solution of the ordinary differential equation $\dot{S} = L(x, \dot{x}, S)$ with the initial data $S(0) = a$.

THEOREM 7. *Let the norms of all second derivatives of* H *and also of the derivatives*

$$\frac{\partial H}{\partial S}, \quad \left.\frac{\partial H}{\partial p}\right|_{x=p=0}, \quad \left.\frac{\partial H}{\partial x}\right|_{p=0}$$

be uniformly bounded. Then for all C *there exists a* t_0 *such that for all smooth convex initial functions* $S_0(x)$ *with the norms of first and second order derivatives bounded by* C *there exists a unique smooth solution of the equation*

$$\frac{\partial S}{\partial t} = H\left(x, \frac{\partial S}{\partial x}, S\right) = 0 \tag{3.2}$$

for $t \leqslant t_0$ *with initial data* S_0. *This solution* $S(t, x)$ *is given by the formula*

$$S(t, x) = \inf_\xi S(t, x, \xi, S_0(x)).$$

This fact follows from the standard method of characteristics. As a consequence we see that the solving operator of the Cauchy problem for equation (3.2) is an additive operator on smooth convex functions. Therefore, we can define the generalized solution of (3.2) by extending the solving operator by continuity and additivity on the set of all functions bounded from below. The correctness theorem follows immediately from this definition (as in the case of Hamilton-Jacobi equation in [5, 6]). For smooth initial functions this definition gives the same result as the method of viscosity solutions.

§4. Homogeneous operators and turnpikes for antagonistic multistep games

Now let us show where homogeneous operators appear from. For simplicity, we restrict ourselves here to the case of a finite state space $X = \{1, \dots, n\}$.

Let us define an antagonistic game on X. Let $p_{ij}(\alpha, \beta)$ denote the probability of transition from the state i to the state j, if the players choose their strategies α and β respectively (α, β belong to some fixed metric spaces with the metric ρ), and $b_{ij}(\alpha, \beta)$ denotes the income of the first player from this transition. The game is said to be a *game with a value*, if

$$\min_{\alpha} \max_{\beta} \sum_{j=1}^{n} p_{ij}(\alpha, \beta)(h^j + b_{ij}(\alpha, \beta))$$

$$= \max_{\beta} \min_{\beta} \sum_{j=1}^{n} p_{ij}(\alpha, \beta)(h^j + b_{ij}(\alpha, \beta)) \tag{4.1}$$

for all $h \in \mathbb{R}^n$. In that case, the operator $B: \mathbb{R}^n \to \mathbb{R}^n$ such that $B_i(h)$ is equal to (4.1) is called the *Bellman operator* of the game. By the dynamic programming method [15] it can be shown that the value of the k-step game defined by the initial position i and the terminal income $h \in \mathbb{R}^n$ of the first player exists and is equal to $B_i^k(h)$.

It is clear that the operator B has the following two properties:

$$B(ae + h) = ae + B(h) \qquad \forall a \in \mathbb{R},\ h \in \mathbb{R}^n,\ e = (1, \dots, 1) \in \mathbb{R}^n, \tag{4.2}$$

$$\|B(h) - B(g)\| \leqslant \|h - g\| \qquad \forall h, g \in \mathbb{R}^n, \tag{4.3}$$

where $\|h\| = \max |h^i|$.

It is interesting that these two properties characterize the Bellman operator of the game.

THEOREM 8. *For each map* $B: \mathbb{R}^n \to \mathbb{R}^n$ *satisfying* (4.2), (4.3) *there exist functions* $p_{ij}(\alpha, \beta)$ *and* $b_{ij}(\alpha, \beta)$ (α, β *belong to some metric spaces*) *such that*

$$p_{ij}(\alpha, \beta) \geqslant 0, \qquad \sum_{j=1}^{n} p_{ij}(\alpha, \beta) = 1,$$

(4.1) *holds and* $B_i(h)$ *is equal to* (4.1).

PROOF. It follows from (4.3) that B_i has partial derivatives almost everywhere such that $\sum_{j=1}^n |\partial B_i/\partial h^j| \leqslant 1$. Then (4.3) implies $\sum_{j=1}^n (\partial B_i/\partial h^j) = 1$. Hence the gradients of B_i belong to the unit simplex

$$\Pi = \left\{ h = (h^1, \dots, h^n) \in \mathbb{R}^n : h^i \geqslant 0,\ \sum_{j=1}^{n} h^j = 1 \right\}.$$

Now let us represent the function B_i in standard game form [16]:

$$B_i(h) = \min_{\alpha\in\mathbb{R}^n}\max_{\beta\in\mathbb{R}^n}(B_i(\alpha) + (F_i(\alpha, \beta), h-\alpha))$$
$$= \max_{\beta\in\mathbb{R}^n}\min_{\alpha\in\mathbb{R}^n}(B_i(\alpha) + (F_i(\alpha, \beta), h-\alpha)),$$

where

$$F_i(\alpha, \beta) = \int_0^1 \operatorname{grad} B_i(\alpha + t(\beta-\alpha))\,dt.$$

Due to the properties of the gradient of B mentioned above, the vector $F_i(\alpha, \beta)$ belongs to Π.

Now let us point out that the most important examples of games with values are games with a finite number of (pure) strategies α, β, when the value exists in so-called mixed strategies. The sets of mixed strategies in that case coincide with some unit simplex in $\mathbb{R}^n$. It follows from the proof of Theorem 8 that the set of corresponding Bellman operators is a dense subset in the set of all operators with properties (4.2), (4.3).

Now we define the quotient space Φ of the space $\mathbb{R}^n$ by the one-dimensional subspace generated by the vector $e = (1, \dots, 1)$. Let $\Pi\colon \mathbb{R}^n \to \Phi$ be the natural projection. The quotient norm in Φ is evidently defined by the formula

$$\|\Pi(h)\| = \inf_{a\in\mathbb{R}}\|h + ae\| = (1/2)\Big(\max_j h^j - \min_j h^j\Big).$$

It is clear that Π has a unique isometric (but not linear) section $S\colon \Phi \to \mathbb{R}^n$. The image $S(\Phi)$ consists of all $h \in \mathbb{R}^n$ such that $\max_j h^j = -\min_j h^j$. Due to the properties (4.2), (4.3) of the map B, the continuous quotient map $\widetilde{B}\colon \Phi \to \Phi$ is defined.

To formulate the main result of the section, we need some additional properties of the transition probabilities:

$$\exists\delta > 0: \quad \forall i, j, \alpha\,\exists\beta: \; p_{ij}(\alpha, \beta) \geqslant \delta, \tag{4.4}$$

$$\exists\delta > 0: \quad \forall i, j\,\exists m: \forall\alpha, \beta, \; p_{im}(\alpha, \beta) > \delta, \; p_{jm}(\alpha, \beta) > \delta. \tag{4.5}$$

Let all $|b_{ij}(\alpha, \beta)|$ be bounded by some constant C.

LEMMA. A) *If* (4.4) *holds and* $\delta < 1/n$, *then* $\widetilde{B}$ *maps each ball of radius* $R \geqslant C\delta^{-1}$ *with center at the origin into itself.* B) *If* (4.5) *holds, then*

$$\|\widetilde{B}(H) - \widetilde{B}(G)\| \leqslant (1-\delta)\|H - G\| \qquad \forall H, G \in \Phi.$$

PROOF. We shall prove only A), because the proof of B) is similar. It follows from the definition that

$$B_i(h) - B_m(h) \leqslant \sum_{j=1}^n p_{ij}(\alpha_1, \beta_1)(b_{ij}(\alpha_1, \beta_1) + h^j)$$
$$- \sum_{j=1}^n p_{mj}(\alpha_0, \beta)(b_{ij}(\alpha_0, \beta) + h^j),$$

where α_1, β_1, and α_0 depend on i and h, while β is arbitrary. Hence

$$B_i(h) - B_m(h) \leqslant 2C + \|h\| \sum_{j=1}^{n} |p_{ij}(\alpha_1, \beta_1) - p_{mj}(\alpha_0, \beta)|.$$

Now let us choose j_0 in such a way that $p_{ij_0}(\alpha_1, \beta_1) > \delta$. Using condition A), we shall choose β so that $p_{ij}(\alpha_0, \beta) > \delta$. Then

$$\sum_{j=1}^{n} |p_{ij}(\alpha_1, \beta_1) - p_{mj}(\alpha_0, \beta)| \leqslant |p_{ij_0} - p_{mj_0}| + (1 - p_{ij_0}) + (1 - p_{mj_0}) \leqslant 2(1-\delta).$$

Consequently,

$$B_i(h) - B_m(h) \leqslant 2C + 2(1-\delta).$$

By definition of the norm in Φ, we get $\|\Pi(B(h))\| \leqslant C + \|h\|(1-\delta)$. Thus, for $h = S(H)$ we obtain $\|\widetilde{B}(H)\| \leqslant C + \|H\|(1-\delta)$. It follows that the map $\widetilde{B}$ takes the ball of radius R to itself, if $C + R(1-\delta) \leqslant R$, or equivalently, $R \geqslant C\delta^{-1}$.

THEOREM 9. A) *If* (4.4) *holds, then there exist a unique* $a \in \mathbb{R}$ *and a vector* $h \in \mathbb{R}^n$ *such that*

$$B(h) = a + h \tag{4.6}$$

and for all $g \in \mathbb{R}^n$ *we have*

$$\|B^m g - ma\| \leqslant \|h\| + \|h - g\|, \tag{4.7}$$

$$\lim_{m\to\infty} B^m g/m = a. \tag{4.8}$$

B) *If* (4.5) *holds, then* h *is unique* (*up to equivalence*) *and*

$$\lim_{m\to\infty} S \circ \Pi(B^m(g)) = S \circ \Pi(h) \quad \textit{for all } g \in \mathbb{R}^n. \tag{4.9}$$

PROOF. It follows immediately from the lemma and fixed-point theorems.

As a consequence we see that the equilibria $\tilde{\alpha}_i$, $\tilde{\beta}_i$ in (4.1), where h is a solution of (4.6), define stationary strategies in the infinite time game. Theorem 9 also implies turnpike theorems for the game considered.

THEOREM 10. *Let* (4.4) *hold. Then for all* $\varepsilon > 0$ *and* $\Omega > 0$ *there exists an* $M \in \mathbb{N}$ *such that if* $\{\alpha(i,t)\}$, $\{\beta(i,t)\}$ *are equilibrium strategies in the* $T > M$ *step game with the terminal income of the first player defined by the vector* g: $\|\Pi(g)\| \leqslant \Omega$, *then for all* $t < T - M$

$$\rho(\alpha(t,i), \tilde{\alpha}_i) < \varepsilon, \qquad \rho(\beta(t,i), \widetilde{\beta}_i) < \varepsilon.$$

PROOF. It follows immediately from (4.9).

Let $\tilde{\alpha}_i$, $\tilde{\beta}_i$ be defined uniquely. Let $Q^\star = (q_1^\star, \dots, q_n^\star)$ denote the stationary distribution for the stationary Markov chain defined on the state space X by these strategies. As in the case of the Markov decision process [11], we now get the turnpike theorem for the state space.

THEOREM 11. *For all* $\varepsilon > 0$, $\Omega > 0$ *there exists an* $M \in \mathbb{N}$ *such that for each* $T > 2M$ *step game with the terminal income* $g \in \mathbb{R}^n$, $\|\Pi(g)\| < \Omega$, *for the first player, we have*

$$\|Q(t) - Q^\star\| < \varepsilon, \quad \text{where } Q(t) = (q_1(t), \dots, q_n(t))$$

and $q_i(t)$ *is the probability of the process being at* $i \in X$ *at time* t, *if the game is played according to the equilibrium strategies.*

REMARK. The natural generalization of conditions (4.4), (4.5) is the case when these conditions hold for some degree of the operator B. This leads to the definition of the so-called cyclic games [17]. It is also possible to extend the results of this section to more general state spaces X.

REFERENCES

1. V. N. Kolokoltsov, *Semiring analogs of linear equivalent spaces*, Abstr. Conf. on Optimal Control, Geometry, Calculus, Kemerovo, 1988, p. 30. (Russian)

2. ———, *Maslov's arithmetic in general topology*, Geometry, Topology and Applications, Moskov. Instrument. Inst., 1990, pp. 64–68. (Russian)

3. S. Yu. Yakovenko, *On the concept of infinite extremal in stationary problems of dynamic optimization*, Dokl. Akad. Nauk SSSR **306** (1989), no. 4, 798–802; English transl. in Soviet Math. Dokl. **49** (1990).

4. V. P. Maslov, *On a new superposition principle for optimization problems*, Uspekhi Mat. Nauk **42** (1987), no. 3, 39–48; English transl. in Russian Math. Surveys **42** (1987).

5. V. N. Kolokoltsov and V. P. Maslov, *The Cauchy problem for the homogeneous Bellman equation*, Dokl. Akad. Nauk SSSR **296** (1987), no. 4; English transl., Soviet Math. Dokl. **36** (1988), no. 2, 326–330.

6. ———, *Idempotent calculus as the apparatus of optimization theory*. I, Funktsional. Anal. i Prilozhen. **23** (1989), no. 1, 1–14; II, **23**, no. 4, 53–62; English transl. in Functional Anal. Appl. **23** (1989).

7. S. M. Avdoshin, V. V. Belov, and V. P. Maslov, *Mathematical aspects of the parallel program synthesis*, Moscow Institute of Electronic Engineering, Moscow, 1984. (Russian)

8. V. P. Maslov and K. A. Volosov (eds.), *Mathematical aspects of computer engineering*, "Mir", Moscow, 1988.

9. J. A. Bather, *Optimal decision procedures for finite Markov chains*, Adv. Appl. Probab. **5** (1973), 328–339, 521–553.

10. I. V. Romanovskiĭ, *Asymptotic behavior of a discrete deterministic process with continuous state space*, Optimal. Planirovanie **8** (1967), "Nauka", Novosibirsk, 171–193. (Russian)

11. V. T. Kolokoltsov, *Turnpikes and infinite extremals in the Markov processes of decision making*, Mat. Zametki **46** (1989), no. 4, 118–120; English transl. in Math. Notes **46** (1989).

12. V. N. Kolokoltsov and V. P. Maslov, *The general form of endomorphisms in the space of continuous functions with values in a numerical semiring with idempotent addition*, Dokl. Akad. Nauk SSSR **295** (1987), no. 2, 283–287; English transl., Soviet Math. Dokl. **36** (1988), no. 1, 55–59.

13. I. V. Romanovskiĭ, *Turnpike theorems for semi-Markov decision processes*, Trudy Mat. Inst. Steklov. **111** (1970), 208–223; English transl. in Proc. Steklov. Inst. Math. **111** (1972).

14. P. I. Dudnikov and S. N. Samborskiĭ, *Endomorphisms of semimodules over semirings with an idempotent operation*, Preprint No. 87-48, Inst. Mat. Akad. Nauk Ukrain., Kiev, 1987. (Russian)

15. R. Bellman, *Dynamic programming*, Princeton Univ. Press, Princeton, NJ, 1960.

16. L. C. Evans, *Some min-max methods for the Hamilton-Jacobi equations*, Indiana Univ. Math. J. **33** (1984), no. 1.

17. T. Parthasarathy and T. Raghavan, *Some topics in two-person games*, American Elsevier Publishing Co., New York, 1971.

18. V. N. Kolokoltsov, *Stochastic Bellman equation as a nonlinear equation in Maslov's space. Perturbation theory*, Dokl. Akad. Nauk SSSR **323** (1992), no. 2, 223–228; English transl. in Soviet Math. Dokl. (to appear).

19. V. N. Kolokoltsov and V. P. Maslov, *Differential Bellman equation and the Pontryagin maximum principle for multicriterial optimization problems*, Dokl. Akad. Nauk SSSR **324** (1992), no. 1, 29–34; English transl. in Soviet Math. Dokl. (to appear).

20. S. Yu. Dobrokhotov, V. N. Kolokoltsov, and V.P. Maslov, *Quantization of the Bellman equation, exponential asymptotics, and tunneling*, this volume.

CENTER FOR OPTIMIZATION AND MATHEMATICAL MODELING, INSTITUTE OF NEW TECHNOLOGIES, KIROVOGRADSKAYA UL., 11, 113587, MOSCOW, RUSSIA

Translated by THE AUTHOR

ADVANCES IN SOVIET MATHEMATICS
Volume 13, 1992

Spectra of Compact Endomorphisms

S. A. LESIN AND S. N. SAMBORSKIĬ

§1. Introduction

Theorems asserting that the spectra of an arbitrary endomorphism A which acts in a finitely generated free semimodule K^n over an idempotent semiring K is nonempty (i.e., theorems about the existence for the endomorphism A of elements $\lambda \in K$ and $f \in K^n$ $(f \not\equiv \mathbf{0})$ such that $Af = \lambda \odot f$) are among the fundamental results of idempotent "linear algebra".

In the simplest case when the semiring K is

$$\mathscr{R} = \{\mathbb{R}^1 \cup (-\infty), \ \oplus = \max, \ \odot = +\},$$

([1], Example 2.1) such a result was first obtained by I. V. Romanovskiĭ [2] and was the consequence of elementary topological properties of the semiring $\mathscr{R}$; namely, properties that allow the use of classical fixed-point theorems.

In [1, 3–5] the semiring K is assumed to satisfy conditions which guarantee the existence of eigenelements, and a description of such conditions of purely algebraic character is given.

In [3–5] the spectra of endomorphisms acting in the semimodule of continuous $\mathscr{R}$-valued functions on a compact set X (i.e., of "integral operators" in the sense of idempotent integration) is investigated and its nonemptiness is established. To prove the existence of the eigenfunction, in these papers the apparatus of nonstandard analysis is applied to the corresponding results obtained in the case of finitely generated semimodules.

Soon afterwards the importance of theorems of this type for applications to mathematical economics (for instance, for the so-called turnpike theorems), in game theory, and in other branches was realized. On the other hand, due to these applications, new proofs of the theorems appeared [6, 7], including proofs of those with weaker restrictions on the "kernels" of the integral operators [7].

In most applications mentioned above, compact endomorphisms arise as a natural class of endomorphisms. In this paper the existence of "eigenvalues" of compact endomorphisms acting in the semimodule of bounded functions

1991 *Mathematics Subject Classification.* Primary 16Y20, 47B38.

over the semiring $\mathscr{R}$ is established in a more general situation than the one considered in the papers [3, 5, 6] (Theorem 2.1). One can prove this fact by reformulating nonstandard modifications of the proof from [3, 5] into the usual language of analysis. Subsequently the case of endomorphisms having more than one eigenvalue (i.e., endomorphisms with a spectrum that is not reduced to a single point) is considered. To determine the spectral sets, we consider the equation

$$Ay \oplus f = \lambda \odot y$$

and study its solvability with respect to y depending on f and on the support of the function f. (The dependence is a specific characteristic connected with the idempotent structures.) If the operation $\oplus$ is invertible (as is ordinary addition), we can immediately reformulate the problem in terms of the properties of the classical resolvent $R_\lambda = (A - \lambda f)^{-1}$.

On the other hand, generalizing "finite-dimensional" results from the papers [1, 4], we give a definition of the spectra in other terms, having an obvious interpretation in dynamic programming problems, in which the kernel $a(x, y)$ of the integral endomorphism is regarded as the "cost" of passing from the point x to the point y of the phase space.

We show the spectra of any compact endomorphism to consist only of "eigenvalues" and to have at most one limit point, and only $\mathbf{0}$, i.e. the neutral element with respect to the operation $\oplus$, can be such a point. Up to a possible infinite multiplicity of the "eigenvalues" this result looks very much like the classical Riesz-Schauder theorem.

Finally, we use the spectral functions to obtain two-sided estimates of the form

$$c_1 \odot \lambda_Y^n(x) \leqslant (A^n f)(x) \leqslant c_2 \odot \lambda_Y^n(x),$$

where A^n are iterates of the compact endomorphism A acting in the semimodule of bounded functions of x defined on a compact set X, Y is the support of the function f, $\lambda_Y(x)$ is the spectral function of the endomorphism A corresponding to the set Y, and $\lambda^n = \lambda \odot \lambda \odot \cdots \odot \lambda$.

The estimates for the Bellman function in dynamic programming problems form one of the evident applications of these inequalities. In the simplest case when the spectrum of the corresponding endomorphism with kernel ("cost") consists of one eigenvalue λ and $\operatorname{supp} f = X$ (in this case, $\lambda_Y(x) \equiv \lambda$) such estimates were obtained earlier by I. V. Romanovskiĭ [8].

§2. Existence of eigenvalues

Let K be an idempotent semiring. Assume further that the semiring K is isomorphic to a closed subsemiring of the semiring $\mathscr{R}$ and introduce on $\mathscr{R}$ a class of metrics which induce the corresponding metrics on K.

Suppose $\varphi(\cdot)\colon \mathbb{R}^1 \to \mathbb{R}^1$ is a continuous function satisfying the following conditions:

(1) $\varphi(\cdot)$ is strictly monotone increasing;

(2) $\varphi(r)$ has a limit equal to zero as $r \to -\infty$;
(3) there exists a constant L such that for all $a, b \in \mathbb{R}^1$ and $h < 0$ we have the inequality

$$|\varphi(a+h) - \varphi(b+h)| \leqslant L|\varphi(a) - \varphi(b)|.$$

Any concave differentiable function satisfying condition (2) (for example, $x \to e^x$) also satisfies conditions (1), (3).

On the semiring $\mathscr{R}$ we now introduce the metric ρ by the formula

$$\rho(a, b) = |\varphi(a) - \varphi(b)|.$$

Then $\mathscr{R}$ (and, consequently, K with the metric induced by the isomorphism which identifies K with a closed subsemiring of $\mathscr{R}$) becomes a complete metric space with a metric continuous in all the variables.

PROPOSITION 2.1. *The metric ρ defined above possesses the following properties*:

(1) *for any set of indices A,*

$$\rho\Big(\sup_{\alpha\in A} a_\alpha, \sup_{\alpha\in A} b_\alpha\Big) \leqslant \sup_{\alpha\in A} \rho(a_\alpha, b_\alpha);$$

(2) *the inequality $\rho(a \odot h, b \odot h) \leqslant L\rho(a, b)$ is valid for all $h \leqslant 1$;*
(3) *convergence in the metric ρ coincides with the convergence in K determined by the order* (*see* [1]).

The proof can be obtained without any difficulty, so we shall omit it.

Suppose X is a totally bounded metric space (i.e., for every $\varepsilon > 0$ there exists a finite covering of the space X by balls of radius ε).

Now consider the semimodule $C(X, K)$ of continuous bounded functions on X with values in the metric semiring K and an "integral operator with kernel a":

$$(Af)(x) = \bigoplus\int_X a(x, y) \odot f(y)\, dy = \sup_{y\in X}(a(x, y) \odot f(y)),$$

where $a\colon X \times X \to K$. In the case when the element

$$\bigoplus\int_X a(x, x)\, dx = \sup_X a(x, x)$$

of the semiring K exists, we call it the *trace* of the operator A and denote it by $\operatorname{Tr} A$.

PROPOSITION 2.2. *Suppose the kernel $a\colon X \times X \to K$ of the integral operator A is a bounded function uniformly continuous in the first argument and equicontinuous in the second one. Then A is a continuous endomorphism of the semimodule $C(X, K)$.*

PROOF. Let $f \in C(X, K)$. One can assume, without loss of generality, that $\|f\| = \sup_X f(x) = 1$. Using Proposition 2.1, we have:

$$\begin{aligned}\rho((Af)(x), (Af)(x')) &= \rho\Big(\sup_{y\in X}(a(x, y) \odot f(y)), \sup_{y\in X}(a(x', y) \odot f(y))\Big) \\ &\leqslant \sup_{y\in X}(\rho(a(x, y) \odot f(y)), a(x', y) \odot f(y)) \\ &\leqslant L \sup_{y\in X} \rho(a(x, y), a(x', y)) \\ &= L\rho(a(x, \cdot), a(x', \cdot)).\end{aligned}$$

The last expression tends to zero as $x \to x'$. Thus, Af is continuous. In a similar way one can show the endomorphism A also to be continuous. Therefore, the proposition is proved.

THEOREM 2.1. *Suppose X is a totally bounded metric space and the kernel $a: X \times X \to K$ of the endomorphism A is a bounded function uniformly continuous in the first argument and equicontinuous in the second one. Then there exist a subsemimodule J of the semimodule $C(X, K)$ $(J \neq 0)$ and an element $\lambda \in K$ such that*

$$(Af)(x) = \lambda \odot f(x)$$

for all $f \in J$, and the maximal element among such λ (with respect to the linear order in K) can be obtained by the formula

$$\lambda = \sup_{i\in N} \lambda(i),$$

where $\lambda(i)$ is the unique root of the equation $\mu^i = \operatorname{Tr} A^i$.

PROOF. Choose a monotonic sequence $\varepsilon_i \to 0$ $(\varepsilon_i > 0)$, and for every positive integer i let E_i be the finite set of centers of balls of radii ε_i covering the space X. We shall assume that $E_i \subset E_j$ for $i < j$ and n_i is the number of elements in the set E_i. In K^{n_i} consider the endomorphism A_i given in the canonical basis in K^{n_i} by the matrix

$$(k(x, y) \mid x, y \in E_i).$$

Let $\tilde{\lambda}_i$ be the eigenvalue of the endomorphism A_i calculated by the formula $\tilde{\lambda}_i = \max_j \mu^j_{i,j}$, where $\mu^j_{i,j} = \operatorname{Tr} A^j_i$ (see [1, 3, 5]). Then the sequence $\{\tilde{\lambda}_i\}_{i=1}^{\infty}$ is monotone increasing (by virtue of the choice of the set E_i and of the formula for $\tilde{\lambda}_i$) and is bounded. So there exists a limit of this sequence as $i \to \infty$. We shall denote it by λ. For every i choose some eigenfunction $f_i: E_i \to K$ corresponding to $\tilde{\lambda}_i$, i.e.,

$$(A_i f_i)(x) = \lambda_i \odot f_i(x), \qquad x \in E_i,$$

and let $\|f_i\| = \sup_{x\in E_i} f_i(x) = 1$.

Now for every $x \in X$ and every sequence $\{x_i\}_{i=1}^{\infty}$ converging to x in X, where $x_i \in E_i$, define the following element $\varphi(x, x_i)$ from K (an upper limit):

$$\varphi(x, x_1) = \overline{\lim_{i\to\infty}} f_i(x_i),$$

which exists because f_i is bounded by the unit of the semiring. Also consider the following function:

$$\psi(x, x_i) = \overline{\lim_{i\to\infty}}((A_i f_i)(x_i)),$$

which exists due to the boundedness of f_i and the assumptions imposed on the kernel a. This function has the following properties:

(1) its values depend only on x and are independent of the choice of the sequence $\{x_i\}_{i=1}^{\infty}$ converging to x $(x_i \in E_i)$. Denote these values by $\psi(x)$;

(2) the function $x \to \psi(x)$ from X to K is continuous.

Indeed, let $x_i \to x$ and $x_i' \to x'$ $(x_i, x_i' \in E_i)$. Then by virtue of the assertion we have:

$$\begin{aligned}\rho(\psi(x, x_i), \psi(x', x_i')) &= \rho\Big(\overline{\lim_{i\to\infty}}(A_i f_i)(x_i), \overline{\lim_{i\to\infty}}(A_i f_i)(x_i')\Big)\\ &= \rho\Big(\overline{\lim_{i\to\infty}} \sup_{y\in E_i}(a(x_i, y)\odot f_i(y)), \overline{\lim_{i\to\infty}} \sup_{y\in E_i}(a(x_i', y)\odot f_i)(y)\Big)\\ &\leqslant \overline{\lim_{i\to\infty}} \sup_{y\in E_i} \rho(a(x_i, y)\odot f_i(y), a(x_i', y)\odot f_i(y)).\end{aligned}$$

The last expression (by virtue of Proposition 2.1 and the assumptions about the kernel a of the endomorphism A) tends to zero as $x \to x'$ independently of the choice of the sequences x_i and x_i'. Moreover, we have

$$\begin{aligned}\psi(x) &= \overline{\lim_{i\to\infty}}(A_i f_i)(x_i) = \overline{\lim_{i\to\infty}} \tilde{\lambda}_i \odot f_i(x_i)\\ &= \overline{\lim_{i\to\infty}} \tilde{\lambda}_i \odot \overline{\lim_{i\to\infty}} f_i(x_i) = \lambda \odot \psi(x, x_i)\end{aligned}$$

as $x_i \to x$ $(x_i \in E_i)$.

Since the left-hand side of the equality is independent of the choice of the sequence $x_i \to x$, we see that the right-hand side also has the same property, i.e., $\varphi(x, x_i) = \varphi$. Therefore, the function $x \to \varphi(x)$ is continuous for the same reason.

Using the equality

$$\psi(x) = (A\varphi)(x),$$

which comes from Proposition 2.1, we have the following relation

$$(A\varphi)(x) = \lambda \odot \varphi(x).$$

It remains to show that $\varphi \not\equiv 0$. Note, at first, that the kernel a can be uniformly extended with respect to the second argument to a mapping $\tilde{a}\colon \widetilde{X} \times \widetilde{X} \to K$ continuous in the first one, where $\widetilde{X}$ is the completion of the space X, the completion $\widetilde{X}$ being a compact set by virtue of the boundedness of X. Repeating all the above arguments (with the same choice of the set E_i) for the operator $\widetilde{A}$ in $C(\widetilde{X}, K)$ with kernel $\tilde{a}$, we obtain a function $\tilde{\varphi}$ coinciding with the function φ on X. The function $\tilde{\varphi}$ is continuous on $\widetilde{X}$ and is not equal to $\mathbf{0}$ identically. Indeed, the assumption that the function $\tilde{\varphi}$ equals zero (identically) contradicts the relation $f_i(x_i') = \mathbf{1}$ which is valid for some $x_i' \in E_i$ and for every i in view of the compactness of $\widetilde{X}$. Since the function φ is the restriction of the continuous function $\tilde{\varphi} \not\equiv 0$ to the dense subset X of $\widetilde{X}$, we see that this function is also not equal to zero identically. Hence, the proof is established.

§3. Spectral sets

Consider a metric compact set X and a semiring K with operations $\oplus$, $\odot$ and neutral elements $\mathbf{0}$ ("zero") and $\mathbf{1}$ ("unit"). Suppose the semiring K is endowed with the min-max metric ρ. Let $C(X, K)$ be the semimodule of continuous functions defined on the compact set X with values in the semiring K, and $A(X)$ be the set of compact endomorphisms on $C(X, K)$ which are linear with respect to the semiring K.

THEOREM 3.1 [7]. *Assume that $A \in A(X)$ is a linear integral operator in the semimodule $C(X, K)$ with kernel $a: X \times X \to K$. The operator A is compact if and only if*:

(1) *$a(x, y)$ is upper semicontinuous for all $x, y \in X$;*
(2) *$a(x, y)$ is uniformly continuous in x equicontinuously in y in the semimodule $C(X, K)$.*

REMARK 3.1. The same integral endomorphism A can be given by different kernels $a(\cdot, \cdot)$ which do not coincide on a set of "idempotent measure zero" (as in functional analysis, the kernels of integral operators are given by the corresponding Lebegue measure almost everywhere). In the sequel, we use the following fact: the correspondence "upper semicontinuous kernel" $\to$ "compact endomorphism" is a one-to-one mapping.

Consider projectors onto the eigenspace of a linear compact endomorphism $A \in A(X)$. The following theorem is an extension of the result obtained by S. Yu. Yakovenko for the case of continuous kernels [6] to the case of arbitrary compact endomorphisms.

THEOREM 3.2. *Let $\lambda \neq \mathbf{0}$ be a unique eigenvalue of the operator A and $f(x)$ the eigenfunction of A that equals $\mathbf{0}$ nowhere on X. Then there exists a unique linear compact operator Ω_λ (called* projector*) onto the eigenspace of the operator A that satisfies the following conditions*:

(1) $\Omega_\lambda^2 = \Omega_\lambda$;
(2) $\Omega_\lambda A = A\Omega_\lambda = \lambda \odot \Omega_\lambda$.

PROOF. Suppose $a_n(\cdot, \cdot)$ is the kernel of the operator A^n, i.e. of the nth iterate of the operator A. Now we shall show that the family of functions

$$\{\lambda^{-n} \odot a_n(\cdot, y) \mid n > 0, \ y \in X\}$$

is precompact in $C(X, K)$.

The validity of the estimate

$$\begin{aligned}
&\rho[a_n(x, y), a_n(\bar{x}, y)] \\
&\quad = \rho\Big(\sup_{z\in X} a(x, z) \odot a_{n-1}(z, y), \sup_{z\in X} a(\bar{x}, z) \odot a_{n-1}(z, y)\Big) \\
&\quad \leqslant \sup \rho(a(x, z) \odot a_{n-1}(z, y), a(\bar{x}, z) \odot a_{n-1}(z, y)) \\
&\quad \leqslant \sup \rho(a(x, y), a(\bar{x}, y))
\end{aligned}$$

for all positive integers $n \geqslant 1$ follows from the upper semicontinuity of the functions a_n.

Since the kernel $a(\cdot, \cdot)$ is uniformly continuous in the first argument equicontinuously in the second one, we see that this estimate implies the equicontinuity in $C(X, K)$ of the family of functions

$$\{\lambda^{-n} \odot a_n(\cdot, y) \mid n > 0, \ y \in X\}.$$

To use the Arzelà theorem we shall show the uniform boundedness in n of this family. Since the eigenfunction $f(x)$ belongs to the semimodule $C(X, K)$ and, by assumption, equals $\mathbf{0}$ ("zero") nowhere on the compact set X, we see that there exist nonzero constants C_1 and C_2 $(C_1, C_2 \in K)$ such that

$$C_1 \leqslant f(x) \leqslant C_2.$$

Thus, for any $n > 0$ we have

$$C_2 \geqslant \sup_{x\in X} f(x) = \sup_{x\in X} \lambda^{-n} \odot (A^n f)(x) \geqslant C_1 \odot \sup_{x, y\in X} \lambda^{-n} \odot a_n(x, y).$$

And so for any positive integer n and for any $x, y \in X$ we obtain the inequality $\mathbf{0} \neq C_2 \odot C_1^{-1} \geqslant \lambda^{-n} \odot a_n(x, y)$. Hence, the precompactness in $C(X, K)$ of the family $\{\lambda^{-n} \odot a_n(\cdot, y) \mid n > 0, \ y \in X\}$ is proved.

Now consider the following function:

$$\omega_\lambda(x, y) = \overline{\lim_{n\to\infty}} \lambda^{-n} \odot a_n(x, y).$$

As an upper bound of upper semicontinuous functions, this function is also upper semicontinuous. The precompactness of the family

$$\{\lambda^{-n} \odot a_n(\cdot, y) \mid n > 0, \ y \in X\}$$

implies the uniform continuity of the function $\omega_\lambda(\cdot, y)$ equicontinuously in y.

Therefore by Theorem 3.1 the function $\omega_\lambda(\cdot, \cdot)$ determines an endomorphism (denoted by Ω_λ) from $A(X)$.

It is easy to see that the kernel ω_λ thus constructed implies conditions (1), (2), and the function $(\Omega_\lambda g)(x)$ is the eigenfunction of A for an arbitrary $g \in C(X, K)$.

Now we shall prove the uniqueness of the projector. Assume that for some operator Q with kernel $q(\cdot, \cdot)$ conditions (1), (2) are satisfied. Then from condition (2) and the upper semicontinuity property of the kernel $q(\cdot, \cdot)$ it follows that

$$q(x, y) = \sup_{z \in X} \lambda^{-n} \odot a_n(x, z) \odot q(z, y)$$

is valid for any positive integer $n > 0$. And so for the function $\omega_\lambda(\cdot, \cdot)$ the following equality

$$q(x, y) = \sup_{z \in X} \omega_\lambda(x, z) \odot q(z, y)$$

holds. Therefore from condition (1) for the kernel $q(\cdot, \cdot)$ of the endomorphism Q, we obtain that the operator Q coincides with the operator Ω_λ.

Hence, the proof is complete.

Suppose A is a compact linear endomorphism acting in $C(X, K)$. For $\lambda \in K \setminus \{\mathbf{0}\}$ (the case $\lambda \equiv \mathbf{0}$ will be considered further on) we shall define the *resolvent* R_λ and the *projector* Ω_λ in the following way:

$$R_\lambda = \sup_{n \geqslant 0} \lambda^{-n} \odot A^n,$$

where $\lambda^0 = \mathbf{1}$, A^0 is an identity endomorphism,

$$\Omega_\lambda = \overline{\lim_{n \to \infty}} \lambda^{-n} \odot A^n.$$

By virtue of Theorem 3.1, R_λ and Ω_λ are integral endomorphisms with the kernels

$$r_\lambda(x, y) = \sup_{n \geqslant 0} \lambda^{-n} \odot a^n(x, y), \qquad \omega_\lambda(x, y) = \lim_{n \to \infty} \lambda^{-n} \odot a^n(x, y)$$

respectively, and, in the case when $r_\lambda(x, y)$ and $\omega_\lambda(x, y)$ are bounded, R_λ and Ω_λ determine compact endomorphisms in $C(X, K)$.

The endomorphism R_λ is called a resolvent not only because of the formal analogy with the resolvent in ordinary analysis but also (as will be seen in what follows) because of the close connection between its properties and the solvability of corresponding equations.

Consider the following equations

$$\lambda \odot y = Ay \oplus f, \tag{1}$$

$$\lambda \odot y = Ay, \qquad y, f \in C(X, K). \tag{2}$$

PROPOSITION 3.1. *Equation* (2) *has a solution* $g(x)$ *from* $C(X, K)$ *which is not identically zero if and only if there exists a function* $f(x)$ *from* $C(X, K)$

such that $(\Omega_\lambda f)(x)$ *is bounded and* $\operatorname{supp}(\Omega_\lambda f)(\cdot) \neq \varnothing$, $g(x)$ *coinciding with* $(\Omega_\lambda f)(x)$.

PROOF. a) Let $g \in C(X, K)$, $g \not\equiv \mathbf{0}$ and $g = \lambda^{-1} \odot Ag$. Then for all positive integers n we have

$$g = \lambda^{-n} \odot A^n g.$$

Therefore, we have $g = \Omega_\lambda g$ and $(\Omega_\lambda g)(x) \in C(X, K)$.

b) Conversely, for some function f from $C(X, K)$ let the inequality $\operatorname{supp}(\Omega_\lambda f) \neq \varnothing$ be satisfied and the function $(\Omega_\lambda f)(x)$ be bounded. Then $(\Omega_\lambda f)(x)$ is a nonzero function from $C(X, K)$. Taking into account the projector properties, we get

$$A(\Omega_\lambda f) = \Omega_\lambda f.$$

Hence $g = \Omega_\lambda f$ is a solution of equation (2) belonging to $C(X, K)$ and is not identically zero. Thus, the proposition is proved.

PROPOSITION 3.2. *The function* y *from* $C(X, K)$ *is a solution of equation* (1) *that is not identically zero if and only if it can be represented in the following form*:

$$y = g \oplus R_\lambda f,$$

where $R_\lambda f$ *is bounded and the function* g *is a solution of equation* (2) *belonging to the space* $C(X, K)$.

PROOF. a) For some function y from $C(X, K)$ let the equality

$$y = \lambda^{-1} \odot Ay \oplus f$$

be satisfied. Then $\lambda^{-1} \odot Ay = \lambda^{-2} \odot A^2 y \oplus \lambda^{-1} \odot Af$ and for all positive integers n the equality

$$y = \lambda^n \odot A^n y \oplus \left(\bigoplus_{k=0}^{n} \lambda^k \odot A^k f \right)$$

is satisfied. Using the fact that the operation $\oplus$ is idempotent, we obtain

$$y = \Omega_\lambda y \oplus R_\lambda f.$$

Since $y \in C(X, K)$, we see that both terms on the right-hand side of this equality are bounded and $g = \Omega_\lambda y$ is a solution of equation (2).

b) Conversely, if the function $R_\lambda f$ is bounded, and the function g is a solution of equation (2) not identically $\mathbf{0}$, then the function $y = g \oplus R_\lambda f$ also belongs to the space $C(X, K)$ and its support is nonempty. The substitution of the last expression in equation (1) yields an identity. The proposition is proved.

It is easy to see that the boundedness of the function $R_\lambda f$ depends only on the boundedness of the kernel $r_\lambda(\cdot, \cdot)$ of the resolvent on $X \times Y$, where

$Y = \operatorname{supp} f$. And so equation (1) is (or is not) simultaneously solvable for all functions f having common support Y.

Next, the boundedness of the function $R_\lambda f$ follows from the boundedness of the function $\Omega_\lambda f$. The converse assertion is also true, but then in a number of cases the function $\Omega_\lambda f$ turns out to be identically zero $(\mathbf{0})$. Thus, for any function from $C(Y, K)$, one of the following possibilities occurs:

a) both equations (1) and (2) are not solvable; in this case the kernel ω_λ of the projector Ω_λ is unbounded on $X \times Y$;
b) equation (1) is solvable, and equation (2) is not solvable; in this case the projector Ω_λ equals $\mathbf{0}$ identically;
c) both equations (1) and (2) are solvable; then the kernel of the projector Ω_λ does not equal $\mathbf{0}$ identically and is bounded.

From the definition of Ω_λ it follows that the set of elements $\lambda \neq \mathbf{0}$ for which possibility a) or possibility b) occurs is an open set. Hence, these possibilities continue to be satisfied under any sufficiently small changes of λ. The corresponding values of $\lambda \neq \mathbf{0}$ will be called *regular elements* and the others (which are also nonzero) will be called *spectral elements with respect to* Y. The set of spectral elements is closed in the induced topology in $K \setminus \{\mathbf{0}\}$.

DEFINITION 3.1. The union Λ of all spectral elements with respect to all possible supports Y will be called the *spectral set* of the *endomorphism* $A \in A(X)$.

Since for positive spectral values of λ the kernel of the projector Ω_λ is bounded and is not identically zero, we see that by Proposition 3.1 there exists a solution of equation $\lambda \odot g = Ag$ which is not identically zero. Thus, the spectral set Λ consists of eigenvalues of the operator A.

THEOREM 3.3. *The spectral set of a compact endomorphism $A \in A(X)$ is either finite or countable with a unique limit point $\mathbf{0}$.*

PROOF. Suppose Λ is the spectral set of a compact endomorphism $A \in A(X)$. For a fixed $x \in X$ let there exist a $\lambda \in \Lambda$ such that $\omega_\lambda(x, x) = c > \mathbf{0}$. Then this λ is unique. Indeed, otherwise there would exist an eigenvalue λ_2 that differs from $\mathbf{0}$ and from λ and for which the equality

$$\omega_{\lambda_2}(x, x) = \overline{\lim_{n\to\infty}}\, a_n(x, x) \odot \lambda_2^{-n} = c \odot \overline{\lim_{n\to\infty}}\, (\lambda^{-1} \odot \lambda_2)^{-n}$$

is satisfied. Since $\lambda^{-1} \odot \lambda_2 \neq \mathbf{1}$, we see that either $\omega_{\lambda_2}(x, x)$ equals $\mathbf{0}$ or the expression under the limit sign on the right-hand side of this equality tends to infinity. Therefore we have a contradiction.

It remains to note that Theorem 2.1 guarantees the existence of $\lambda_{\max}$, where $\lambda_{\max}$ is the maximal eigenvalue of the endomorphism A (i.e., for all $\lambda \in \Lambda$ we have $\lambda_{\max} \geqslant \lambda$). This means that the inequality

$$\sup_{x\in X} \omega_{\lambda_{\max}}(x, x) < \infty$$

holds. Thus the function $S: X \to \Lambda$ is well defined by the rule:

$$S(x) = \begin{cases} \mathbf{0} & \text{if } \omega_\lambda(x, x) = \mathbf{0} \text{ for all } \lambda > \mathbf{0}, \\ \lambda & \text{if } x \in \operatorname{supp} f(\cdot), \text{ where } f(y) = \omega_\lambda(y, y) \text{ for all } y \in X. \end{cases}$$

It is easy to see that this function is upper semicontinuous. Now we shall show that this function is also constant on any open set where it is continuous. Let $\bar{x}$ be a continuity point of the function S and $S(\bar{x}) = \lambda \neq \mathbf{0}$. Assume that there exists a sequence $\{x_i\}_{i=0}^{\infty}$ converging to the point $\bar{x}$ and satisfying the following conditions:

(1) $S(x_i) \neq S(x_j)$, if $i \neq j$;
(2) $S(x_i) = \lambda_i \xrightarrow[i\to\infty]{} S(\bar{x}) = \lambda \neq \mathbf{0}$.

One can assume that there exists a number i_0 such that $S(x_i) > \mathbf{0}$ for all numbers $i > i_0$. For every x_i with $i > i_0$ there exists a number n_i such that $a_{n_i}(x_i, x_i) > \mathbf{0}$. Fix a number $i > i_0$ and let δ be the distance between the point x_i and the point belonging to $\{x_i\}_{i=0}^{\infty}$ which is nearest (in the metric of the compact set X) to x_i. Let this nearest point have the number k. Taking into account the equicontinuity of the family $a_n(\cdot, y)$, we see that the connected components of $\operatorname{supp} b_{n_i}(\cdot, x_i)$ and $\operatorname{supp} b_{n_k}(\cdot, x_k)$ that contain the points x_i and x_k, respectively, cannot contain balls of radius δ with centers at the points x_i and x_k. In the opposite case the points x_i and x_k would belong to the same support of the kernel $\omega_{\bar{\lambda}}(y, y)$ for some $\bar{\lambda}$, i.e., the equality $S(x_i) = S(x_k)$ which is at variance with the choice of the sequence $\{x_i\}_{i=0}^{\infty}$ would be satisfied. Since the sequence $\{x_i\}_{i=0}^{\infty}$ is Cauchy and the function $S(x)$ is continuous at the point $\bar{x}$, we obtain $S(\bar{x}) = \mathbf{0}$. Thus if condition (1) holds for some sequence $\{x_i\}_{i=0}^{\infty}$ converging to the continuity point $\bar{x}$ of the function S, then condition (2) does not hold for this sequence. And so the function $S(x)$ is constant on any open subset of X which is the domain of continuity of this function, and the sequence $\{S(x_i)\}_{i=0}^{\infty}$ either stabilizes or has $\mathbf{0}$ as its limit. Now the proof is straightforward from the definition of S. Thus the theorem is proved.

§4. Spectral functions and estimates of iterates of compact endomorphisms

We shall suppose, as before, that X is a compact set, A is a compact endomorphism, and Ω_λ is the projector onto its eigensubsemimodule corresponding to some eigenvalue $\lambda > \mathbf{0}$ from the spectral set of the endomorphism A.

Let $x \in X$ and Y be some subset of the set X. Denote by $L(x, Y)$ the set of eigenvalues λ of the endomorphism A (i.e., the set of elements from $\mathbb{R}^1$) for which the inequality

$$\mathbf{0} < \sup_{y \in Y} \omega_\lambda(x, y) < \infty$$

is valid. In the case when the equality

$$\sup_{y\in Y} \omega_\lambda(x, y) = \mathbf{0}$$

holds for all $\lambda > 0$, the set $L(x, Y)$ will be assumed to coincide with the one-point set $\{\mathbf{0}\}$.

ASSERTION 4.1. *For every fixed* $x \in X$, $Y \subset X$, *the set* $L(x, Y)$ *is exactly a one-point set.*

PROOF. 1) We shall show that for every $x \in X$ and $Y \subset X$ the set $L(x, Y)$ is nonempty.

Consider the sets

$$\overline{\alpha} = \left\{\alpha \in \mathbb{R}^1 \mid \forall c \in \mathbb{R}^1, \sup_{y\in Y} \omega_\alpha(x, y) > c\right\}$$

and

$$\underline{\alpha} = \left\{\alpha \in \mathbb{R}^1 \mid \sup_{y\in Y} \omega_\alpha(x, y) = \mathbf{0}\right\}.$$

Let $\lambda \in \overline{\alpha}$. We show that λ does not belong to the closure of $\underline{\alpha}$ in $\mathbb{R}^1$. On the contrary, assume that $\lambda = \lim_{i\to\infty} \lambda_i$, where $\lambda_i \in \underline{\alpha}$. Then

$$\begin{aligned}
\sup_{y\in Y} \omega_\lambda(x, y) &= \sup_{y\in Y} \overline{\lim_{n\to\infty}}\, a_n(x, y) \odot \lambda^{-n} \\
&= \sup_{y\in Y} \overline{\lim_{n\to\infty}}\, a_n(x, y) \odot \lambda_i^{-n} \odot (\lambda_i^{-1} \odot \lambda)^{-n} \\
&\leqslant \overline{\lim_{n\to\infty}} \lim_{i\to\infty} \sup_{y\in Y} a_n(x, y) \odot \lambda_i^{-n} \odot (\lambda_i^{-1} \odot \lambda)^{-n} \\
&= \overline{\lim_{n\to\infty}} \lim_{i\to\infty} \mathbf{0} \odot (\lambda_i^{-1} \odot \lambda)^{-n} = \mathbf{0},
\end{aligned}$$

which contradicts the choice of $\lambda \in \overline{\alpha}$.

Similarly, one can show that elements of $\underline{\alpha}$ do not belong to the closure of the set $\overline{\alpha}$. And so for any $x \in X$ and $y \subset X$ there exists a $\lambda \in \mathbb{R}^1$ such that $\mathbf{0} \subset \sup \omega_\lambda(x, y) < \infty$.

According to Definition 3.1, λ belongs to the set Λ. The last statement proves that the set $L(X, Y)$ is not empty.

2) Now we prove that for every pair (x, Y) such that $L(x, Y) \neq \{\mathbf{0}\}$ the set $L(x, Y)$ consists of a single point $\lambda \in \Lambda$.

We argue by *reductio ad absurdum*. Assume that there exist two values λ_1 and λ_2 $(\lambda_1 > \lambda_2 > \mathbf{0})$ for which the following inequalities

$$\mathbf{0} < \sup_{y\in Y} \omega_{\lambda_2}(x, y) = c_1 < \infty,$$

$$\mathbf{0} < \sup_{y\in Y} \omega_{\lambda_1}(x, y) = \sup_{y\in Y} \overline{\lim_{n\to\infty}}\, a_n(x, y) \odot \lambda_1^{-n} \leqslant c_1 \odot \overline{\lim_{n\to\infty}} (\lambda_1 \odot \lambda_2^{-1})^{-n}$$

are valid. Since the sequence $\{(\lambda_1 \odot \lambda_2^{-1})^{-n}\}_{n=1}^{\infty}$ tends to $\mathbf{0}$ as $n \to \infty$, we get a contradiction. Thus, the proof is complete.

Assertion 4.1 implies that the function $\Lambda_Y(\cdot)\colon X \to \Lambda \cup \{\mathbf{0}\}$ is well defined by the following rule:

$$\Lambda_Y(x) = \begin{cases} \lambda & \text{if } \mathbf{0} < \sup_{y \in Y} \omega_\lambda(x, y) < \infty, \\ \mathbf{0} & \text{if } \sup_{y \in Y} \omega_\lambda(x, y) = \mathbf{0} \text{ for all } \lambda > \mathbf{0} \end{cases}$$

for all $x \in X$. The function $\Lambda_Y(x)$ will be called the *spectral function of the endomorphism A corresponding to the subset Y.*

Consider the following equivalence relation on the set X : x is equivalent to y, if $x = y$ or there exists $\lambda \in \Lambda$ such that $\mathbf{0} < \omega_\lambda(x, y) \odot \omega_\lambda(y, x) < \infty$. Let $\{X_i\}_{i \in I}$ be the set of such equivalence classes. To every such class corresponds a unique λ from the set Λ. Note that the classes defined by the condition $x = y$ are one-point sets and the value $\mathbf{0}$ can correspond only to these classes, whenever $a_n(x, x) = \mathbf{0}$ for all positive integers $n > \mathbf{0}$. It is easy to see that the equivalence relation ensures the validity of the relation

$$a_k(x, y) \odot a_n(y, x) = \mathbf{0}$$

for all integers $n, k > 0$ and any x, y belonging to different classes. By using this property one can define a partial order Γ on the set $\{X_i\}_{i \in I}$ according to the rule:

$$(X_i, X_j) \in \Gamma \iff \{\exists x \in X_i, y \in X_j \mid a(x, y) > \mathbf{0}\}.$$

Thus, the collection $(\{X_i\}_{i \in I}, \Lambda \cup \{\mathbf{0}\}, \Gamma)$ defines an oriented graph without cycles and with weights $\lambda \in \Lambda \cup \{\mathbf{0}\}$ at the vertices X_i $(i \in I)$.

DEFINITION 4.1. The *spectrum* Spec A *of an arbitrary operator A from $A(X)$* is defined as the oriented graph $(\{X_i\}_{i \in I}, \Lambda \cup \{\mathbf{0}\}, \Gamma)$.

EXAMPLE 4.1. Consider the segment $X = [0, 2] \subset \mathbb{R}^1$ and a compact endomorphism A acting in the semimodule of bounded functions on X over the semiring $(\mathbb{R}^1_+, \max, \cdot)$. Suppose that the endomorphism A has the kernel $a(x, y)$ of the following form:

$$a(x, y) = \begin{cases} 2x, & y = 0, \\ 1 - (x-1)^2, & y \in (0, 2), \\ 2(2 - x), & y = 2. \end{cases}$$

The compact set X can be divided into two equivalence classes: the open interval $X_1 = (0, 2)$ and the two-point set $X_2 = \{0, 2\}$. The weights corresponding to these classes are also determined without any difficulty as follows:

$$\lambda_1 = a(1, 1) = 1, \qquad \lambda_2 = \sqrt{\operatorname{Tr} A^2} = \sqrt{a(2, 0) \circ a(0, 2)} = 4.$$

The values λ_1 and λ_2 are the eigenvalues of the endomorphism A. The spectrum of this operator can be represented as the following oriented graph with two vertices:

$$(\lambda_1) \bullet \longrightarrow \bullet (\lambda_2)$$

THEOREM 4.1. *Suppose X is a compact metric space, A is a compact endomorphism acting in the semimodule $C(X, \mathscr{R})$. Then for any continuous function $f: X \to \mathscr{R}$ with support Y there exists a constant $\bar{c} \neq \mathbf{0}$ such that for all positive integers n and for all $x \in \operatorname{supp} \Lambda_Y(\cdot)$ the following estimate*

$$(A^n f)(x) \leqslant \bar{c} \odot \Lambda_Y^n(x) \tag{3}$$

is satisfied, where $\Lambda_Y(\cdot)$ is the spectral function of the endomorphism A defined above.

PROOF. Fix a point $x \in \operatorname{supp} \Lambda_Y$ and let $\lambda = \Lambda_Y(x) \in \mathbb{R}^1$. By the definition of spectral elements, the kernel $\omega_\lambda(\cdot, \cdot)$ of the projector Ω_λ and, therefore, the kernel $r_\lambda(\cdot, \cdot)$ of the resolvent R_λ, are bounded on the set $\operatorname{supp} \Lambda_Y \times X$. It is equivalent to the existence of a constant $\bar{c} \neq \mathbf{0}$ such that the following inequalities

$$\bar{c} \geqslant (R_\lambda f)(x) \geqslant \lambda^{-n} \odot (A^n f)(x)$$

take place for all positive integers n, whence we get the estimate (3) (for all positive integers n and for any $x \in \operatorname{supp} \Lambda_Y$). The fact that the set X is compact guarantees the existence of the constant $\bar{c}$ which is the same for all $x \in X$. Thus, the proof is over.

THEOREM 4.2. *Let X be a metric compact set, A a compact endomorphism acting in the semimodule $C(X, \mathscr{R})$, $f: X \to \mathscr{R}$ a bounded function with support Y. Assume that there exists a positive integer S such that the condition*

$$\operatorname{supp} A^S f = \operatorname{supp} A^{S+1} f$$

holds. Then for every $x \in X$ there exists a number $\underline{c} > \mathbf{0}$ such that for all positive integers n the following estimate

$$(A^{S+n} f)(x) \geqslant \underline{c} \odot \Lambda_Y^{S+n}(x)$$

is valid, where $\Lambda_Y(\cdot)$ is the spectral function of the endomorphism A.

PROOF. Note, at first, that if the condition

$$\operatorname{supp} A^S f = \operatorname{supp} A^{S+1} f$$

holds for some S, then it remains valid when we substitute any positive integer $S' > S$ for S. Indeed, denote by Θ the support of the function $A^S f$. By virtue of the assumptions, we have $\sup_{\xi \in \Theta} a(x, \xi) = \mathbf{0}$ for $x \notin \Theta$ and for every $x \in \Theta$ there exists a point $\xi^0 \in \Theta$ such that $a(x, \xi^0) > \mathbf{0}$. Therefore, if for some positive integer S' we have $\operatorname{supp}(A^{S'} f) = \Theta$, then for $S' + 1$ we obtain:

$$(A^{S'+1} f)(x) = \sup_{\xi \in \Theta} a(x, \xi) \odot (A^{S'} f)(\xi) = \mathbf{0} \quad \text{for } x \notin \Theta,$$

$$(A^{S'+1} f)(x) \geqslant a(x, \xi^0) \odot (A^{S'} f)(\xi^0) > \mathbf{0} \quad \text{for } x \in \Theta,$$

i.e., $\operatorname{supp}(A^{S'+1} f) = \Theta$.

Fix a point $z \in X$. If $\Lambda_Y(z) = \mathbf{0}$, then the theorem is true for all $\underline{c}$. Let $\Lambda_Y(z) = \lambda \neq \mathbf{0}$. Then the definition of the spectral function implies the inequality

$$\mathbf{0} < \sup_{y \in Y} \omega_\lambda(z, y) < \infty,$$

where ω_λ is the kernel of the projector Ω_λ. Since the function $f(\cdot)$ is continuous on the compact set X and the function $\omega_\lambda(\cdot, \cdot)$ is upper semi-continuous on $X \times X$, we see that there exists a $y \in X$ such that

$$(\Omega_\lambda f)(z) = \sup_\xi \omega_\lambda(z, \xi) \odot f(\xi) = \omega_\lambda(z, y) \odot f(y). \tag{4}$$

Fix this y to the end of the proof. Consider the sequence

$$\{z_i \mid i = 0, \dots, \infty, \ z_i \in X\}$$

such that $z_0 = z$ and for all positive integers i we have the inclusions $z_i \in \arg\sup_\xi a_i(z, \xi) \odot \omega(\xi, y)$ which can be written as equalities:

$$\omega_\lambda(z, y) \odot \lambda^i = a_i(z, z_i) \odot \omega_\lambda(z_i, y). \tag{5}$$

Since $\operatorname{supp}(A^n f)$ equals Θ for any positive integer $n > S$, we see that the inclusion $\operatorname{supp} \omega_\lambda(\cdot, y) \subset \Theta$ holds. Thus $z_i \in \Theta$ for any number $i \geqslant 0$. Let $\bar{z}$ be a limit point of the sequence $\{z_i\}_{i=0}^\infty$. The continuity property of the function $\omega_\lambda(\cdot, y)$ yields

$$\omega_\lambda(z, y) = \omega_\lambda(z, \bar{z}) \odot \omega_\lambda(\bar{z}, y),$$

which means that any limit point of the sequence $\{z_i\}_{i=0}^\infty$ also belongs to the set Θ.

Denote by M_z the closure of the subset $\{z_i \mid i = 0, \dots, \infty\}$ in the set X. The previous inclusion implies $M_z \subset \Theta$.

Since the functions $(A^S f)(\cdot)$ and $\omega_\lambda(\cdot, y)$ do not vanish on the set M_z and this set is a closed subset of the compact set X, we see that:

(1) there exists a constant $c_1 \neq \mathbf{0}$ such that for all $x \in M_z$ we have the following inequalities

$$(A^S f)(x) > c_1, \qquad \omega_\lambda(x, y) > c_1; \tag{6}$$

(2) there exists a constant $c_2 \neq \mathbf{0}$ such that for all $x \in M_z$ the following inequalities

$$(A^S f)(x) \geqslant c_2 \odot \lambda^S \odot \omega_\lambda(x, y) \odot f(x) > \mathbf{0} \tag{7}$$

hold. (The last inequality comes from the boundedness from above of the function $\omega_\lambda(\cdot, y)$ on the compact set M_z.)

Recall that $\Lambda_Y(z) = \lambda$, i.e., $(\Omega_\lambda f)(z) \neq \mathbf{0}$. For any positive integer n, the inclusion $z_n \in \Theta$ implies the inequality

$$(A^{n+S} f)(z) = (A^n(A^S f))(z) \geqslant a_n(z, z_n) \odot (A^S f)(z_n).$$

From this inequality, using (6) and (7), we obtain

$$(A^{n+S}f)(z) \geqslant a_n(z, z_n) \odot \omega_\lambda(z_n, y) \odot c_2 \odot \lambda \odot f(y).$$

Finally, using (4) and (5) and denoting $(\Omega_\lambda f)(z) \odot c_2 \neq \mathbf{0}$ by $\underline{c}$, we obtain the inequality

$$(A^{n+S}f)(z) \geqslant \lambda^{n+S} \odot (\Omega_\lambda f)(z) \odot c_2 = \Lambda_Y^{n+S}(z) \odot \underline{c}$$

which completes the proof.

REMARK 4.1. The restriction (determined by the condition of the theorem) of the function f to the support is essential for the lower estimate. To see this consider the endomorphism from Example 3.1.

Let

$$f(x) = \begin{cases} 1 - x & \text{for } x \in [0, 1), \\ 0 & \text{for } x \in [1, 2]. \end{cases}$$

Then $\operatorname{supp} f = [0, 1)$.

Consider the iterates of the operator A at the point $x = 0$. At this point the spectral function $\Lambda_{[0,1)}(0)$ takes the value 4, i.e., at this point $\Lambda_{[0,1)}(0)$ differs from $\mathbf{0}$. But the condition

$$\operatorname{supp} A^S f = \operatorname{supp} A^{S+1} f$$

from the theorem does not hold: for any positive integer n the expression $(A^n f)(0) \cdot 4^{-n}$ equals 0, if n is odd, and equals 1, if n is even.

It is obvious that the lower estimate does not exist in this case.

REFERENCES

1. P. I. Dudnikov and S. N. Samborskiĭ, *Endomorphisms of finitely generated free semimodules*, this volume.

2. I. V. Romanovskiĭ, *Optimization of stationary control of discrete deterministic process in dynamic programming*, Kibernetika (Kiev) **2** (1967), 66–78; English transl. in Cybernetics **3** (1967).

3. P. I. Dudnikov and S. N. Samborskiĭ, *Spectra of endomorphisms of semimodules over semirings with an idempotent operation*, Dokl. Akad. Nauk SSSR **308** (1989), no. 4, 363–366; English transl. in Soviet Math. Dokl. **40** (1990).

4. ______, *Endomorphisms of semimodules over semirings with an idempotent operation*, Preprint No. 87-48, Inst. Mat. Akad. Nauk Ukrain., Kiev, 1987. (Russian)

5. ______, *Endomorphisms of semimodules over semirings with an idempotent operation*, Izv. Akad. Nauk SSSR Ser. Mat. **55** (1991), 93–109; English transl. in Math. USSR-Izv. **38** (1992).

6. S. Yu. Yakovenko, *On the concept of infinite extremal in stationary problems of dynamic optimization*, Dokl. Akad. Nauk SSSR **306** (1989), no. 4, 798–802; English transl. in Soviet Math. Dokl. **40** (1990).

7. V. N. Kolokoltsov, *Linear additive and homogeneous operators in idempotent analysis, and their applications*, this volume.

8. I. V. Romanovskiĭ, *Asymptotic behavior of a discrete deterministic process with continuous state space*, Optimal. Planirovanie **8** (1967), "Nauka", Novosibirsk, 171–193. (Russian)

CENTER FOR OPTIMIZATION AND MATHEMATICAL MODELING, INSTITUTE OF NEW TECHNOLOGIES, KIROVOGRADSKAYA UL., 11, 113587, MOSCOW, RUSSIA

Translated by I. PETRENKO

ADVANCES IN SOVIET MATHEMATICS
Volume 13, 1992

Stationary Hamilton-Jacobi and Bellman Equations (Existence and Uniqueness of Solutions)

V. P. MASLOV AND S. N. SAMBORSKIĬ

1. For first order nonlinear partial differential equations, specifically, for Hamilton-Jacobi and Bellman equations of the form

$$H(x, Dy) = 0, \tag{1}$$

where D is the gradient of a scalar function y of a variable x belonging to a domain X of Euclidean space, many problems of mathematical physics, optimal control theory, game theory reduce to boundary value problems, for example, to the Dirichlet problem: $y = y_0$ on the boundary Γ of the domain X. As is well known, the solutions of such problems, which usually have a clear physical meaning either corresponding to optimal controls or giving the strategies of games, are not differentiable as a rule.

The simplest example of such kind is the equation

$$\|Dy(x)\| = 1,$$

where x belongs to a bounded domain $X \subset \mathbb{R}^n$ with boundary $\Gamma = \bar{X} \setminus X$ ($\bar{X}$ is the closure of X in $\mathbb{R}^n$, $\|\cdot\|$ is the norm in $\mathbb{R}^n$), with the boundary value condition $y|_\Gamma = 0$. The function assigning to every point x in $\bar{X}$ the minimal distance from x to Γ satisfies this equation and the boundary value condition at the points of differentiability. This function is globally continuous but not differentiable. The use of "generalized differentiation" in Sobolev spaces does not solve the problem, since in this case one cannot expect the solutions to be unique: an infinite number of functions satisfy the equation and boundary value conditions. And so different versions of the definition of solutions of Hamilton-Jacobi or Bellman type equations have been suggested (see [1–6]). In the papers [2–4] and others, the solutions are defined as functions for which sub- and super-differentials satisfy a system of one-sided inequalities (this leads to the consideration of multivalued mappings). In another approach used in game theory [5], the solution is defined as a function satisfying a system of min-max inequalities connected with the

1991 *Mathematics Subject Classification*. Primary 35F20, 49L05; Secondary 16Y60, 49N60.

initial equation. The definition of the solution as a limit of solutions of the Dirichlet problem for the singularly perturbed partial differential equation of second order

$$\varepsilon\Delta y + H(x, Dy) = 0$$

is also physical. In spite of all the advantages and the validity of these approaches, the desire to say, following the classical tradition, that a solution of the equation is an object such that its substitution into the equation makes sense and reduces the equation to the identity (in some sense) of its two parts remains strong.

In this paper we show how the ideas of idempotent analysis allow this desire to be fulfilled along the lines of the classical weak solution approach. The distinction from the classical scheme of distributions consists, first of all, in using another algebraic structure in the set of real numbers, namely, the structure of an idempotent semiring. We define function spaces whose elements are no more "generalized" objects than, say, elements of the space L^p and are defined as well as these elements as equivalence classes with respect to an equality "almost everywhere", only the measure being different. The classical approach of Sobolev space and derivative in the sense of districtions works well for linear equations, if one applies, as usual, the linear structure that comes from the fields $\mathbb{R}$ or $\mathbb{C}$. The same approach is used to analyze nonlinear equations as well, but in this case it is necessary to use another algebraic structure (in the case considered, it is necessary to use the structure of an idempotent semiring) in $\mathbb{R}$. Undoubtedly, the choice of this structure depends on the form of the equations. For equations of the form (1) whose solutions reflect extremum principles (as in the example given above) the use of algebraic structures with idempotent operations of max or min type is completely natural, so that the principles of superposition are expressed in terms of these operations [7, 8].

Similarly to the Sobolev space, we define the space of distributions, or "generally differentiable functions", and "generalized differentiation". Differential operators act in these spaces and the solutions of the corresponding differential equations are realized in the very traditional sense. Besides the above-mentioned concept of solutions as classes of equivalent functions (with a new equivalence relation) the continuity condition for differential operators as a basic requirement on Hamiltonians H generating these operators is also new. Such a view is especially promising for the further development of perturbation theory which is important in many applications.

2. Let $\mathscr{R}$ denote the set $\mathbb{R}\cup(+\infty)$ with the operations $a\oplus b = \min(a, b)$, $a\odot b = a+b$, and the neutral elements $\mathbf{0} = (+\infty)$ and $\mathbf{1} = (0)$, respectively. Let $\mathscr{B}(\bar{X}, \mathscr{R})$ be the semimodule of bounded mappings from $\bar{X}$ to $\mathscr{R}$ (remark that the boundedness of f for the operation $\oplus = \min$ means the existence of a number $k \in \mathbb{R}$ such that $f(x) \geqslant k$ for all $x \in X$). As in [6–8], we now define a "scalar product" on $\mathscr{B}(\bar{X}, \mathscr{R})$ by the formula

$$(f, g) = \bigoplus \int_{\bar{X}} f(x) \odot g(x)\, dx = \inf_{x \in \bar{X}} (f(x) + g(x)).$$

Define the set Φ of test functions as the collection of functions that are continuously differentiable in the domain X and tend to $+\infty$ as the argument tends to the boundary Γ of the domain $\bar{X}$. The main definition for the sequel is the following one.

DEFINITION 1 (equivalence). Two functions f and g from $\mathscr{B}(\bar{X}, \mathscr{R})$ are *equivalent* $(f \sim g)$, if $(f, \varphi) = (g, \varphi)$ for any function $\varphi \in \Phi$.

The scalar product $(\cdot, \cdot)$ allows us to introduce the following weak convergence.

DEFINITION 2 (weak convergence). A sequence of functions (equivalence classes) $\{f_i\}_{i=1}^{\infty}$ *weakly converges* to a function $f \in \mathscr{B}(\bar{X}, \mathscr{R})$ (we write $f = w_0\text{-}\lim f_i$), if for all functions φ from Φ we have

$$\lim_{i \to \infty} (f_i, \varphi) = (f, \varphi).$$

In each equivalence class f there exists a unique representative that is a lower semicontinuous function; we denote it by $\bar{f}$. If $\bar{f}$ is bounded from above, then the equivalence class $-f$ can be defined uniquely for f as the class containing the function $-\bar{f}$.

DEFINITION 3 (the space $P^0(\bar{X}, \mathscr{R})$ of scalar functions on $\bar{X}$). An element of the space $P^0(\bar{X}, \mathscr{R})$ is defined as a set of equivalence classes f (of functions from $F(\bar{X}, \mathscr{R})$) such that $-f$ exists and $f = -(-f)$; the convergence is weak.

Since any two different continuous functions belong to different equivalence classes, the following natural imbedding is defined (and continuous)

$$i_0 \colon C(\bar{X}, \mathbb{R}) \to P^0(\bar{X}, \mathscr{R}).$$

Let ξ be a vector field on $\bar{X}$, and U a differentiable coordinate neighborhood with coordinates $(x_1, \dots, x_n)$. The principal part in U of the field ξ is given by the collection $(\xi_1, \dots, \xi_n)$ of scalar functions $\xi \colon U \to \mathscr{R}$.

DEFINITION 4 (equivalent vector fields). Two vector fields ξ and η on the set $\bar{X}$ are *equivalent*, if for any local coordinate neighborhood U the components ξ_i and η_i of the principal parts in U of the vectors ξ and η are equivalent for each $i = 1, \dots, n$.

DEFINITION 5 (the space $P^0(T\bar{X})$ of vector fields on $\bar{X}$). An element of the space $P^0(T\bar{X})$ is defined as an equivalence class of vector fields for which in every local coordinate neighborhood U the components of the principal parts belong to $P^0(\bar{U}, \mathscr{R})$. The convergence is defined componentwise in $P^0(\bar{U}, \mathscr{R})$ in every coordinate neighborhood U.

REMARK 1. Consider the sequence of the functions $f_n = (1/n) \sin nx$ on some interval I in $\mathbb{R}$. This sequence converges in $P^0(\bar{I}, \mathscr{R})$ to the function 0. The derivatives $f_n' = \cos nx$ converge in $P^0(\bar{I}, \mathscr{R})$ to the function (equivalence class) identically equal to -1. But the sequence of vector fields on I that are the gradients Df_n is not convergent in $P^0(TI)$. Indeed, in the coordinate system on I induced by the standard one in $\mathbb{R}$, the principal part of the gradient Df_n coincides with $f_n' = \cos nx$ and converges to the function $x \to -1$, which is the principal part of the vector field. In the coordinate system in I given by the mapping $x \to -x$ from I to $\mathbb{R}$ (change of direction in $\mathbb{R}$) the principal parts of Df_n, which are equal to $-\cos nx$, converge to the function $x \to -1$ which is the principal part of the vector field $-\xi$ (but not of ξ, since in this coordinate system the field ξ has the principal part $x \to 1$). This example illustrates the natural requirement that the convergence of vector fields be independent of the choice of a coordinate system. When the coordinate system changes, the limits of the principal parts of convergent sequences must change according to the transformation rules for the principal parts of vector fields, i.e., the Jacobi matrix of the coordinate transformation must be applied to them. It is useful to take into account this remark in the next definition.

DEFINITION 6 (weak P^1-convergence of scalar functions). A sequence of functions f_i (differentiable on X) P^1-*converges* to the function (equivalence class) $f \in \mathscr{B}(X, \mathscr{R})$ (we shall write $f = w_1\text{-}\lim f_i$) if $f = w_0\text{-}\lim_{i\to\infty} f_i$, and the vector field $w_0\text{-}\lim Df_i$ exists and belongs to $P^0(TX)$.

PROPOSITION 1. *If for two sequences f_i and g_i we have $w_1\text{-}\lim f_i = w_1\text{-}\lim g_i = f$, then $w_0\text{-}\lim Df_i = w_0\text{-}\lim Dg_i$, i.e., the $w_0\text{-}\lim$ of the gradients depends only on f but not on the sequence which approximates f.*

PROOF. First let the function f be differentiable. Then, whenever $F = w_1\text{-}\lim Df_i$ and $F \neq (\partial f/\partial x_1, \partial f/\partial x_2, \dots, \partial f/\partial x_n)$, on some open set U for some index j we have $F_j(x) \neq (\partial f/\partial x_j)(x)$ (for $x \in U$), and the function $F_j - \partial f/\partial x_j$ does not change sign on this set; this contradicts the assumption $w_0\text{-}\lim f_i = f$.

In the general case, having assumed that for some j there exist two limits F_j and G_j, $F_j \neq G_j$, of the sequences $\partial f_i/\partial x_j$ and $\partial g_i/\partial x_j$, respectively, and $\partial f_i/\partial x_j < \partial g_i/\partial x_j$, one can choose a differentiable function φ such that

$$\frac{\partial f}{\partial x_j} < \frac{\partial \varphi}{\partial x_j} < \frac{\partial g}{\partial x_j}$$

on some open set U. This immediately leads to a contradiction in the same way as in the case of a differentiable function f.

By virtue of Proposition 1, for functions (equivalence classes) that are P^1-limits of sequences of differentiable functions the vector field, called the *gradient* Df of the function f, is uniquely defined (as class of equivalent vector fields).

DEFINITION 7 (the space $P^1(\bar{X}, \mathscr{R})$ of scalar functions). A function f (equivalence class) belongs to the space $P^1(\bar{X}, \mathscr{R})$, if it is a w_1-limit of a sequence of differentiable functions. The convergence in the semiring $P^1(\bar{X}, \mathscr{R})$ is determined by P^1-convergence in the sense of Definition 6, where the operation D can be applied to any elements from $P^1(\bar{X}, \mathscr{R})$.

ASSERTION 1. *In each equivalence class $\{f\}$ from the space $P^1(\bar{X}, \mathscr{R})$ there is a unique continuous function, which will be denoted by f.*

The proof is obvious.

In the sense of this assertion we shall say that the elements from the space $P^1(\bar{X}, \mathscr{R})$ are continuous functions. In particular, restrictions to the boundary Γ of the domain $\bar{X}$ are defined and continuous; this is important in what follows.

It is clear that any differentiable function $f: \bar{X} \to \mathbb{R}$ determines an equivalence class in $P^1(\bar{X}, \mathscr{R})$, so that there exists a unique continuous imbedding

$$i_1: C^1(\bar{X}, \mathbb{R}) \to P^1(\bar{X}, \mathscr{R})$$

of the space of differentiable functions on $\bar{X}$ in $P^1(\bar{X}, \mathscr{R})$.

ASSERTION 2. *The following diagram*

$$\begin{array}{ccc} C^1(\bar{X}, \mathbb{R}) & \xrightarrow{i_1} & P^1(\bar{X}, \mathscr{R}) \\ {\scriptstyle D}\downarrow & & {\scriptstyle D}\downarrow \\ C(TX) & \xrightarrow{i_0} & P^0(TX) \end{array}$$

is commutative, where $C(TX)$ is the space of continuous vector fields with uniform convergence, and the operators D denote the ordinary gradient (on the left) or the gradient D defined above (on the right) of a function from $P^1(\bar{X}, \mathscr{R})$.

The proof is evident. (Note that the mapping $D: P^1 \to P^0$ is continuous.)

The following assertion is useful in the sequel.

ASSERTION 3. *Let $\{f_i\}$ $(i \geqslant 1)$ be a sequence of differentiable functions which is w_0-convergent to the function $f \in P^0(\bar{X}, \mathscr{R})$, and in some coordinate neighborhood U, $(x_1, \dots, x_n)$, for every $j = 1, \dots, n$ the sequence of the functions $\partial f_i/\partial x_j$ w_0-converges to $F_j \in P^0(\bar{X}, \mathscr{R})$, and the sequence of the functions $-\partial f_i/\partial x_j$ w_0-converges to $G_j \in P^0(\bar{X}, \mathscr{R})$. For the sequence $\{f_i\}$ to w_1-converge to f in U it is necessary and sufficient to have $F_j = -G_j$ for every $j = 1, \dots, n$ (in the sense of elements from P^0, i.e., as equivalence classes).*

PROOF. The necessity is evident. To prove the sufficiency, we note that if for each of the two w_0-convergent sequences a_i and b_i the upper and lower semicontinuous enveloping curves of the limits determine the same equivalence classes (i.e., elements from $P^0(\bar{X}, \mathscr{R})$), then any linear combination

$\{\lambda_1 a_i + \lambda_2 b_i\}$ $(\lambda_1, \lambda_2 \in \mathbb{R})$ also possesses the same property. And so the application of the Jacobi matrix of the change of coordinates map transforms the limits according to the transformation law for the principal part of vector fields, which proves the assertion.

The space $P^0(\bar{X}, \mathscr{R})$ is a topological semimodule over the semiring $\mathscr{R}$, i.e., for any functions f_1 and f_2 from $P^0(\bar{X}, \mathscr{R})$ (equivalence classes), and $\lambda \in \mathscr{R}$, the functions $f_1 \oplus f_2$ and $\lambda \odot f_1$ belong to P^0 (since these functions are defined pointwise on representatives of equivalence classes, it can be easily seen that these operations are well defined), and the operation of addition $\oplus$ and multiplication $\odot$ by λ determine continuous mappings $P^0 \times P^0 \to P^0$ and $P^0 \times \mathscr{R} \to P^0$, respectively. (Note, however, that the operation of ordinary addition $+$ is not continuous: $\sin^2 nx \to 0$, $\cos^2 nx \to 0$ as $n \to \infty$, but $\sin^2 nx + \cos^2 nx = 1$.)

The space $P^1(\bar{X}, \mathscr{R})$ is an (algebraic) semimodule over $\mathscr{R}$, i.e., $f \oplus g \in P^1$, and $\lambda \odot f \in P^1$ for $f, g \in P^1$ and $\lambda \in \mathscr{R}$, but the operation $\oplus$ is not continuous, as one can see by means of the following example. Let $f \equiv 0$, and $g_\varepsilon(x) = |x| - \varepsilon$. We have $D(f \oplus g_0) = D(f) \equiv 0$, but

$$w_0\text{-}\lim_{\varepsilon \to 0}(D(f \oplus g_\varepsilon)) \not\equiv 0.$$

PROPOSITION 2. *There exists a subsemimodule in* $P^1(\bar{X}, \mathscr{R})$ *maximal with respect to inclusion containing all differentiable functions (i.e., elements from* $i_1 C^1(\bar{X}, \mathbb{R})$*) in which the operation of addition* $\oplus$ *and multiplication* $\odot$ *by elements from* $\mathscr{R}$ *are continuous in the topology induced from* P^1.

The proof follows from the inner description of elements of the subsemimodule indicated above and given below in Proposition 3.

Denote by $\mathscr{A}(\bar{X}, \mathscr{R})$ (or merely by $\mathscr{A}$) the subsemimodule of the semimodule $P^1(\bar{X}, \mathscr{R})$ mentioned in Proposition 2. Its elements can be characterized by means of the following property.

PROPOSITION 3. *The function* $f \in P^1(\bar{X}, \mathscr{R})$ *belongs to the subsemimodule* $\mathscr{A}$ *if and only if the following property holds: suppose that for some* $x_0 \in X$ *and some differentiable nonnegative function* $\varphi \in i_1 C^1(\bar{X}, \mathbb{R})$ *we have* $(f, \varphi) = f(x_0)$; *then the function* f *is differentiable at the point* x_0 *(here differentiability is understood in the classical sense).*

PROOF. The property mentioned in Proposition 3 can be reformulated in the following way: if φ is a differentiable function such that $f - \varphi$ reaches its minimum at the point x_0, then f is differentiable at the point x_0. Let $f \in \mathscr{A}$, and suppose that for some differentiable function φ the function $f - \varphi$ reaches its minimum at a point x_0. Then the function f is differentiable at the point x_0, since otherwise the sequence $f \oplus \lambda_i \odot \varphi$ would not be w_1-convergent to the function $f \oplus \bar{\lambda} \odot \varphi$, where $\bar{\lambda}$ satisfies $\bar{\lambda} \odot \varphi(x_0) = f(x_0)$ and $\lambda_i \to \lambda$, $\lambda_i > \lambda$.

Conversely, suppose the existence of a differentiable function φ such that $f-\varphi$ reaches its minimum at x_0 implies the differentiability of the function f at the point x_0. Assume that $f \notin \mathscr{A}$; therefore, there exists a sequence φ_i which is w_1-convergent to φ such that $f \oplus \varphi_i$ does not converge to $f \oplus \varphi$ in P^1. One can consider the functions φ_i and φ to be differentiable. This implies the existence of a point x_0 at which the function $f-\varphi$ reaches its minimum; otherwise small P^1-perturbations of the function φ do not remove jump changes of the derivatives of $f \oplus \varphi$. But then the function f is differentiable at the point x_0, and for such a function it is impossible for the functions $f \oplus \varphi_i$ not to converge in P^1 to the function $f \oplus \varphi$ for smooth φ and φ_i.

EXAMPLE 1. The function $x \to |x|$ does not belong to $\mathscr{A}(\bar{I}, \mathscr{R})$, where I is an interval in $\mathbb{R}$ containing 0, but the function $x \to -|x|$ belongs to $\mathscr{A}(\bar{I}, \mathscr{R})$.

3. Let H be a mapping from $\bar{X} \times \mathbb{R}^n$ to $\mathbb{R}$ (the Hamiltonian) which will be assumed continuous. The equations considered have the form (1):

$$H(x, Dy) = 0,$$

where $y \in P^1(\bar{X}, \mathscr{R})$, Dy is the gradient of the function y in the sense described above, and the equality is understood in the sense of an equality in $P^0(\bar{X}, \mathscr{R})$, i.e., the coincidence of equivalence classes. Thus, we consider the mapping from $P^1(\bar{X}, \mathscr{R})$ into $P^0(\bar{X}, \mathscr{R})$ determined by the function H and denoted below by $\mathscr{H}$:

$$\mathscr{H}(y)(x) = H(x, Dy(x)).$$

The basic assumption concerning the mapping $\mathscr{H}$ is in its continuity.

EXAMPLE 2. Let X be an interval in $\mathbb{R}$, and $H(x, p) = -p^2 + a(x)$, where $a \in C(\bar{X}, \mathbb{R})$. The mapping $\mathscr{H}: y \to -(y'_x)^2 + a(x)$ from P^1 to P^0 is continuous. On the other hand, the mapping $\mathscr{H}_1$ determined by the function $H(x, p) = p^2$ is not continuous. Indeed, let $g(x) = -|x|$, and $g_n(\cdot)$ be any sequence of differentiable functions convergent in the topology of P^1 to g, for instance,

$$g_n(x) = \begin{cases} -|x| & \text{for } |x| \geqslant n^{-1}, \\ -nx^2/2 - n^{-1}/2 & \text{for } |x| \leqslant n^{-1}. \end{cases}$$

Then

$$\mathscr{H}(g) = w_0\text{-}\lim \mathscr{H}(g_n) = -1, \quad \mathscr{H}_1(g) = 1,$$

but

$$\mathscr{H}_1(g_n) = \begin{cases} 1 & \text{for } |x| \geqslant n^{-1}, \\ n^2x^2 & \text{for } |x| < n^{-1}, \end{cases}$$

so that $w_0\text{-}\lim \mathscr{H}_1(g_n) \neq 1$.

ASSERTION 4. *Let* $X \subset \mathbb{R}^n$ *and the mapping* $y \to \mathcal{H}(y)$ *from* $P^1(\bar{X}, \mathcal{R})$ *to* $P^0(\bar{X}, \mathcal{R})$, *where* $\mathcal{H}(y)(x) = H(x, Dy(x))$, *be continuous. Then for every* $x \in X$ *and* $\lambda \in \operatorname{Im} H(x, \cdot)$ *the set*

$$\Lambda_{x,\lambda} = \{p \in \mathbb{R}^n \mid H(x, p) \geqslant \lambda\}$$

is a convex subset in $\mathbb{R}^n$.

PROOF. Suppose $a, b \in \mathbb{R}^n$, $x_0 \in X$, and $\langle \cdot, \cdot \rangle$ is the Euclidean scalar product in $\mathbb{R}^n$. Consider the function

$$\varphi_{a,b}(x) = \min\{\langle a, x\rangle + m, \langle b, x\rangle\},$$

where the number $m \in \mathbb{R}$ is chosen so that

$$\langle a, x_0\rangle + m = \langle b, x_0\rangle.$$

The function $\varphi_{a,b}$ belongs to the space $P^1(\bar{X}, \mathcal{R})$, and a sufficiently small neighborhood U of the point x_0 is split by the hyperplane

$$\mathfrak{M}^0 = \{x \mid \langle a - b, x\rangle + m = 0\}$$

into the subsets U^+ where $D\varphi = a$, and U^- where $D\varphi = b$. Let φ_n be a sequence of differentiable functions in $\bar{X}$ which w_1-converges to the function $\varphi_{a,b}$. For any interior point of the segment I in $\mathbb{R}^n$ that connects the points a and b, we have the inclusion $z \in \operatorname{Im} D\varphi_n$ for n sufficiently large.

Let x_0 and λ be such that the set $\Lambda_{x_0,\lambda}$ is nonempty and does not consist of a single point. Assume, on the contrary, that there exist two different points a and b from $\Lambda_{x,\lambda}$ and a point z from the segment I in $\mathbb{R}^n$ joining a and b that does not belong to $\Lambda_{x_0,\lambda}$. Let $l = H(x_0, z)$. Then $l < \lambda$. We have $\mathcal{H}(\varphi_{a,b})(x_0) \geqslant \lambda$, whereas for the terms φ_n of the sequence w_1-convergent to $\varphi_{a,b}$ there always exists a point x_n such that $\mathcal{H}(\varphi_n)(x_n) = l$ holds, x_n converging to x as $n \to \infty$. Thus, the sequence $\mathcal{H}(\varphi_n)$ does not converge in P^0 to $\mathcal{H}(\varphi_{a,b})$ contradicting the continuity of $\mathcal{H}: P^1 \to P^0$. This contradiction proves the convexity of the set $\Lambda_{x_0,\lambda}$.

REMARK 2. Usually the convexity of the Hamiltonian $H(x, \cdot)$ is not necessary for the mapping $\mathcal{H}$ to be continuous. For instance, if $X \subset \mathbb{R}$, and $H: \mathbb{R} \to \mathbb{R}$ does not depend on x, then the mapping $y \to \mathcal{H}y = H(Dy)$ from P^1 into P^0 is continuous, whenever the function H is monotone or has a unique local extremum being maximum.

We shall say that a subset $\Lambda \subset \mathbb{R}^n$ is *strictly convex*, if all the inner points of every segment that connects two different points from Λ are the inner points of Λ.

THEOREM 1. *Suppose that* X *is a bounded open domain in* $\mathbb{R}^n$, $\Gamma = \bar{X} \setminus X$ *is its boundary, and* $H: \bar{X} \times \mathbb{R}^n \to \mathbb{R}$ *is a continuous mapping. Let the following conditions be satisfied:*

(1) *the mapping* $\mathscr{H}: P^1(\bar{X}, \mathscr{R}) \to P^0(\bar{X}, \mathscr{R})$, $(\mathscr{H}y)(x) = H(x, Dy)$, *is continuous;*

(2) *the set* $\Lambda_x = \{p \in \mathbb{R}^n \mid H(x, p) \geqslant 0\}$ *is nonempty, compact, and strictly convex for all* $x \in X$;

(3) *there exists a function* $\bar{y} \in P^1(\bar{X}, \mathscr{R})$ *satisfying the inequality*

$$H(x, D\bar{y}(x)) \geqslant 0. \tag{2}$$

Then equation (1) *has a solution* $y \in \mathscr{A}(\bar{X}, \mathscr{R})$ *such that the restrictions of* y *and* $\bar{y}$ *to the boundary* Γ *coincide.*

PROOF. The restriction of the function $\bar{y}$ to Γ will be denoted by g. Assume that some function F is the pointwise supremum of all P^1-solutions f of the inequality $\mathscr{H}(f) \geqslant 0$ whose restrictions to Γ coincide with the function g, i.e.,

$$F(x) = \sup f(x).$$

The set of such functions f is nonempty because of condition (3) of the theorem. We want to show that the function F is a solution of the equation $\mathscr{H}(y) = 0$ (which may have more than one solution). We shall investigate the properties of the function F in several steps.

ASSERTION 5. *The function* F *is continuous and its restriction to the boundary* Γ *coincides with the function* g.

PROOF. Assume, on the contrary, that the function F is not continuous at some point $\bar{x}$. Let f be a solution of inequality (2) that is close to the function F at the point $\bar{x}$, and $\{f_i\}$ be a sequence of functions from $C^1(\bar{X}, \mathbb{R})$ w_1-approximating f. Then from the definition of the convergence in P^1 for any $K \in \mathbb{R}$ and for some sequence $x_i \to \bar{x}$ the inequality

$$\|Df_i(x_i)\| > K$$

follows. Hence, the gradient Df (in the sense of the definition, given above, of the gradient of a function from P^1, i.e., as an element from $P^0(T\bar{X})$) does not satisfy the inequality $H(x, Df) \geqslant 0$ either (by condition (2) of the theorem).

ASSERTION 6. *The function* F *satisfies the following condition: if for some* $x_0 \in X$ *and a differentiable nonnegative function* $\varphi \in i_1C^1(X, \mathbb{R})$ *we have* $(f, \varphi) = f(x_0)$, *then the function* f *is differentiable at the point* x_0 *(in the classical sense).*

PROOF. Take $\varepsilon > 0$. By definition of the function F as a supremum of P^1-solutions, in its ε^3-neighborhood in the C^1-norm there exists a function G_ε from P^1 possessing the following properties:

a) $G_\varepsilon(x) \leqslant F(x)$ for all $x \in \bar{X}$;

b) the function G_ε satisfies inequality (2);

c) the restriction of the function G_ε to the boundary Γ coincides with g (for the function G_ε we can take, say, the maximum of a finite number of P^1-solutions of inequality (2) that satisfy the boundary value condition g on Γ).

Assume that the function F is not differentiable at the point $x_0 \in X$; then in some neighborhood U of the point x_0 for the function G_ε the following additional property is valid:

d) in the domain $\bigcup_{x\in U} \Lambda x$ there exist two points A and B, $A \neq B$, with neighborhoods $\mathscr{O}_A$ and $\mathscr{O}_B$ such that

$$\{DG_\varepsilon(x) \mid x \in U\} \cap \mathscr{O}_A \neq \varnothing, \qquad \{DG_\varepsilon(x) \mid x \in U\} \cap \mathscr{O}_B \neq \varnothing.$$

To prove Assertion 6 we shall argue by reductio ad absurdum and use the criterion determining elements of the subsemimodule $\mathscr{A}$ given in Proposition 3. Note that the condition $(F, \varphi) = F(x_0)$ for a nonnegative differentiable function φ is equivalent to the existence of a differentiable function ψ such that the function $F - \psi$ reaches its minimum at the point x_0. Thus, let such a function ψ exist, the function F not being differentiable at the point x_0. For ε sufficiently small, the function G_ε (which can be assumed to satisfy conditions a)–d)) is such that $G_\varepsilon - \psi$ has its minimum at some inner point $\bar{x}$ of the neighborhood U.

The set

$$\{C \mid C = D\theta(\bar{x}), \theta(\cdot) \in C^1(\bar{U}, \mathbb{R}), \text{ and the function } G_\varepsilon - \theta \text{ has its minimum at the point } \bar{x}\}$$

is a convex set in $\mathbb{R}^n$ containing points from $\mathscr{O}_A$ and $\mathscr{O}_B$. So one can choose a differentiable function $\bar{\psi}$ so that the function $G_\varepsilon - \bar{\psi}$ has its minimum at the point $\bar{x}$ and, moreover, $D\bar{\psi}(\bar{x})$ belongs to the interior of the set

$$\{z \in \mathbb{R}^n \mid z = \lambda a + (1-\lambda)b, \text{ where } a \in \mathscr{O}_A,\ b \in \mathscr{O}_B \text{ and } \lambda \in [0, 1]\}.$$

Due to the strict convexity of the set $\Lambda_{\bar{x}}$, there exists an ε such that for all x from the ε-neighborhood of the point $\bar{x}$,

$$D\bar{\psi}(x) \in \Lambda_x. \tag{3}$$

Now for $\lambda \in \mathbb{R}$ consider the function

$$G_{\varepsilon,\lambda} = \begin{cases} \max\{\lambda + \bar{\psi}, G_\varepsilon\} & \text{in } U, \\ G_\varepsilon & \text{outside } U. \end{cases}$$

Let λ_0 be a value such that $\lambda_0 + \bar{\psi}(\bar{x}) = G_\varepsilon(\bar{x})$. Then for λ sufficiently close to λ_0 but greater than $\lambda_0 + \varepsilon^3$, the function $G_{\varepsilon,\lambda}$ has the following properties:

a) at some points from U the function $G_{\varepsilon,\lambda}$ has values that exceed values of the function F at these points;

b) $G_{\varepsilon,\lambda}$ belongs to the space $P^1(\bar{X}, \mathscr{R})$;
c) $G_{\varepsilon,\lambda}$ satisfies inequality (2) (by (3));
d) on Γ, the function $G_{\varepsilon,\lambda}$ coincides with the restriction of G_ε to Γ and, therefore, with g.

The existence of the function $G_{\varepsilon,\lambda}$ with properties a)–d) contradicts the fact that the function F is the supremum of all P^1-solutions of inequality (2) and is equal to g on Γ. Hence the function F is differentiable at the point x_0.

ASSERTION 7. *The function F belongs to the space $P^1(\bar{X}, \mathscr{R})$.*

PROOF. Now we shall construct a sequence of P^1-functions that w_1-converges to F. Since F is continuous (because of Assertion 6), we see that for every ball U of radius ε belonging to X there exists a differentiable function φ_U such that $F - \varphi_U$ has its minimum at some inner point $x(U)$ of U. We shall assume that $\varphi_U(x(U)) = F(x(U))$ and take a finite covering $\mathfrak{U}_N$ of the domain X by a family of N neighborhoods U_α $(\alpha = 1, \dots, N)$ of sufficiently small radius ε_N tending to zero as $N \to \infty$. Consider the function

$$G_N(x) = \max_{\alpha'}(\varphi_{U_{\alpha'}}(x)),$$

where the maximum is taken over all α' from 1 to N such that $x \in U_{\alpha'}$. The collection of functions $\{G_N\}$ corresponding to the collection of the coverings $\{\mathfrak{U}_N\}$ possesses the following properties:

a) every function G_N belongs to the space P^1 (as the maximum of a finite number of differentiable functions);
b) the sequence G_N converges to the function F in the space $P^0(\bar{X}, \mathscr{R})$ as $N \to \infty$;
c) in each fixed local coordinate system $(x_1, \dots, x_n)$ in some neighborhood $W \subset X$, the sequence of functions $\partial G_N/\partial x_i$ converges in the space $P^0(\bar{X}, \mathscr{R})$ to some function $G^{W,i}$ belonging to the semiring $P^0(\bar{X}, \mathscr{R})$.

Now we show that the sequence G_N converges in P^1, using Assertion 3. By virtue of this assertion, it is sufficient to show that the upper and lower semicontinuous (respectively, from above and from below) envelopes of the sequence $\partial G_N/\partial x_i$ determine the same equivalence class in $P^0(\bar{X}, \mathscr{R})$. By construction of the function G_N and by Assertion 6, for every neighborhood U and for every $\varepsilon > 0$ there exist a neighborhood $V \subset U$ and a number N_0 such that for every $N > N_0$ at points $x, y \in V$ of differentiability of G_N the inequality

$$\left| \frac{\partial G_N(x)}{\partial x_i} - \frac{\partial G_N(y)}{\partial x_i} \right| < \varepsilon$$

is valid for all $i = 1, \dots, n$.

This implies the required property for the envelopes of $\partial G_N/\partial x_i$, and hence, the fact that $F \in P^1$.

ASSERTION 8. *The gradient DF of the function F satisfies equation* (1).

PROOF. Assume, on the contrary, that $\mathscr{H}(F) = \lambda$, where the function λ (equivalence class) belongs to P^0 and never vanishes. By definition of F, we obtain the inequality $\lambda \geqslant 0$. Then, by definition of F and of the elements from P^0 (as equivalence classes), there exist a nonempty neighborhood U and a number $\lambda_0 > 0$ such that $\lambda(x) > \lambda_0$ for any x from U and any representative $\lambda(\cdot)$ of the equivalence class of λ. Let α be a differentiable function equal to zero outside the neighborhood U and positive at some point x_0 from U. The sequence of functions $F + \mu_i\alpha$ converges in P^1 to F as $\mu_i \to 0$ and $\mu_i > 0$. Because of the continuity of the mapping $\mathscr{H}: P^1 \to P^0$ (condition (1) of the theorem) the sequence $S_i = \mathscr{H}(F + \mu_i\alpha)$ converges in P^0 to λ, and therefore, we have the inequality $S_i \geqslant 0$ for μ_i sufficiently small. But this contradicts the fact that the function F is the supremum of all P^1-solutions of inequality (2) with the boundary value g, since the inequality $F(x) + \mu_i\alpha(x) > F(x)$ is strict in U.

From Assertions 6 and 7 it follows that $F \in \mathscr{A}(\bar{X}, \mathscr{R})$ and from Assertions 5 and 8 it follows that F is a solution of equation (1) satisfying the boundary value condition. Thus, the theorem is proved.

4. In this section we shall study the uniqueness of solutions from the semimodule $\mathscr{A}(\bar{X}, \mathscr{R})$ of the Dirichlet problem for equation (1).

DEFINITION 8 (stable solutions). A solution $y_0 \in \mathscr{A}(\bar{X}, \mathscr{R})$ of equation (1) with the boundary value condition $y_0|_\Gamma = g$ is called *stable*, if there exists a $\lambda_0 < 0$ such that for every $\lambda \in [\lambda_0, 0]$ there is a solution $y_\lambda \in \mathscr{A}(\bar{X}, \mathscr{R})$ of the equation

$$H(x, Dy(x)) = \lambda \tag{4}$$

with boundary value condition $y_\lambda|_\Gamma = g$ such that in the metric of $C(\bar{X}, \mathbb{R})$ we have

$$y_0 = \lim_{\lambda \to 0, \lambda < 0} y_\lambda .$$

The construction method for the solution described in Theorem 1 (as the supremum of P^1-solutions of inequality (2) satisfying the boundary value condition) immediately implies the following proposition.

PROPOSITION 4. *Let condition* (2) *of Theorem* 1 *be replaced by the following condition*:

(2′) *the set $\Lambda_{x,\lambda} = \{p \in \mathbb{R}^n \mid H(x, p) \geqslant \lambda\}$ is nonempty, compact, and strictly convex for all λ from some interval $(\lambda_0, 0]$, where $\lambda_0 < 0$.*

Then there exists a stable solution $y_\lambda \in \mathscr{A}(\bar{X}, \mathscr{R})$ of equation (4) *with the boundary value condition $y_\lambda|_\Gamma = \bar{y}|_\Gamma$ for $\lambda \in (\lambda_0, 0]$.*

REMARK 3. From condition (2) of Theorem 1 it follows that $\Lambda_{x,\lambda}$ is nonempty and is compact for λ from some interval $(\lambda_0, 0]$, where $\lambda_0 < 0$.

THEOREM 2. *Let the mapping*

$$\mathscr{H}: P^1(\bar{X}, \mathscr{R}) \to P^0(\bar{X}, \mathscr{R}), \qquad (\mathscr{H}y)(x) = H(x, Dy(x))$$

be continuous and the equation $\mathscr{H}y = 0$ with boundary value g on Γ possess a stable solution from $\mathscr{A}(\bar{X}, \mathscr{R})$. Then this solution is the unique stable solution of the equation $\mathscr{H}y = 0$ with the boundary value g in the semimodule $A(\bar{X}, \mathscr{R})$.

PROOF. Assume that f_1 is stable and f_2 is any other (not necessarily stable) solution of equation (1) with the boundary value g on Γ. Let h_λ be a solution of equation (4) with boundary value g on Γ that is close (in the metric of C) to the solution f_1 for λ close to 0, $\lambda_0 < \lambda < 0$.

ASSERTION 9. *Suppose that the mapping $\mathscr{H}: P^1(\bar{X}, \mathscr{R}) \to P^0(\bar{X}, \mathscr{R})$ is continuous, a function f from $\mathscr{A}(\bar{X}, \mathscr{R})$ satisfies equation* (1), *and a function h from $\mathscr{A}(\bar{X}, \mathscr{R})$ satisfies equation* (4) *(for $\lambda < 0$) and $f|_\Gamma = h|_\Gamma$. Then for all $x \in X$ we have the inequality $h(x) \geqslant f(x)$.*

PROOF. Assume, on the contrary, that there exists a point $x \in X$ such that $f(x_0) > h(x_0)$. Since both of these functions belong to $P^1(\bar{X}, \mathscr{R})$, we get

$$f(x) > h(x), \qquad x \in U, \tag{5}$$

in some neighborhood U of the point x_0 $(U \subset X)$. Consider the function $S_\mu = f \oplus \mu \odot h = \min\{f(\cdot), \mu + h(\cdot)\}$ belonging to the semimodule $\mathscr{A}(\bar{X}, \mathscr{R})$ for all μ from $\mathbb{R}$. Consider, also, the set

$$\mathscr{O}_\mu = \{x \in X \mid \mu + h(x) < f(x)\}.$$

For $\mu > 0$ this set does not intersect the boundary Γ, since the function S_μ coincides with the function f in some neighborhood of the boundary Γ. For $\mu > 0$ sufficiently small, the set $\mathscr{O}_\mu$ is nonempty in view of inequality (5), and for μ sufficiently large the set $\mathscr{O}_\mu$ is empty. When μ increase, we have the inclusion $\mathscr{O}_{\mu_1} \subset \mathscr{O}_{\mu_2}$ for $\mu_1 > \mu_2$. Therefore, there exists a $\bar{\mu}$ such that for $\mu \geqslant \bar{\mu}$ the function S_μ coincides with f, and for $\mu < \bar{\mu}$ the function S_μ coincides with $\mu \odot h$ on an open set. The sequence $S_i = f \oplus \mu_i \odot h$ of elements from $\mathscr{A}$ converges in P^1 to the function $S = f \oplus \bar{\mu} \odot h = f$ as $\mu_i \to \bar{\mu}$ $(\mu_i < \bar{\mu})$ by definition of the semimodule $\mathscr{A}$. The continuity of the mapping $\mathscr{H}: P^1 \to P^0$ implies the convergence of the sequence $\mathscr{H}(S_i)$ to $\mathscr{H}(S) = \mathscr{H}(f)$. But this is not true, since $\mathscr{H}(S) = 0$, and for every i the function $\mathscr{H}(S_i)$ is equal to $\lambda < 0$ on the open set $\mathscr{O}_{\mu_i}$, hence $w_0\text{-}\lim \mathscr{H}(S_i) \neq 0$. The contradiction obtained proves that $f(x) \leqslant h(x)$ for all $x \in X$.

To prove Theorem 2 it is sufficient to note that the stable solution, according to Assertion 9, always coincides with the supremum of all possible $\mathscr{A}$-solutions with the boundary value g and, thus, is unique.

EXAMPLE 2. Let $\bar{X} = [-1, 1]$, $g(1) = g(-1) = 0$ and $H(x, p) = -p^2 + x^4$. It is obvious that the equation

$$-(Dy)^2 + x^4 = 0 \tag{6}$$

has differentiable (hence belonging to $\mathscr{A}(\bar{X}, \mathscr{R})$) solutions

$$\varphi_1 : x \to |x|^3/3 - 1/3 \quad \text{and} \quad \varphi_2 : x \to -|x|^3/3 + 1/3$$

with zero boundary values. Hence for $\nu_1, \nu_2 \in \mathscr{R}$ such that $\nu_1 \oplus \nu_2 = \mathbf{1}$ the function $\nu_1 \odot \varphi_1 \oplus \nu_2 \odot \varphi_2$ is also a solution of equation (6) belonging to $\mathscr{A}$ and satisfying the boundary value condition. (More generally: all $\mathscr{A}$-solutions of an equation of the form (1) with a fixed boundary value constitute a convex set (in the operations $\oplus$ and $\odot$) [9].) Thus, there is an infinite number of solutions from $\mathscr{A}(\bar{X}, \mathscr{R})$ of the Dirichlet problem for this equation, but only the solution φ_1 is stable. Remark that in the family of equations

$$-(Dy)^2 + x^4 = \lambda$$

for $\lambda < 0$ there exists a unique solution from $\mathscr{A}$ with zero boundary value, and for $\lambda > 0$ there exists no solution at all. This reflects a general regularity: the nonstable solutions of the Dirichlet problem for equation (1) arise only for critical values of parameters, and the uniqueness and stability of the solution to the Dirichlet problem is generic.

REMARK 4. If one regards equation (1) from the point of view of the papers [3–5], then the viscosity solutions (in [3, 4]) or the min-max solutions (in [5]) coincide with those given by Theorems 1 and 2. The $\mathscr{A}$-solutions considered above coincide with limits of solutions of the Dirichlet problem as $\varepsilon \to 0$ for the singularly perturbed equation of the form

$$\varepsilon \Delta y + H(x, Dy) = 0,$$

when such limits exist (here Δ is the Laplace operator).

REFERENCES

1. S. N. Kruzhkov [Kružkov], *Generalized solutions of nonlinear equations of first order with several variables*, Mat. Sb. **70** (1966), no. 3, 394–415. (Russian)

2. P.-L. Lions, *Generalized solutions of Hamilton-Jacobi equations*, Pitman, London, 1982.

3. M. G. Crandall and P.-L. Lions, *Viscosity solutions of Hamilton-Jacobi equations*, Trans. Amer. Mat. Soc. **277** (1983), 1–42.

4. I. Capuzzo-Dolcetta and P.-L. Lions, *Hamilton-Jacobi equations with state constraints*, Trans. Amer. Mat. Soc. **318** (1990), no. 2, 643–685.

5. A. I. Subbotin, *Min-max solutions and Hamilton-Jacobi equations*, "Nauka", Moscow, 1991. (Russian)

6. V. N. Kolokoltsov and V. P. Maslov, *Idempotent calculus as the apparatus of optimization theory*, I, Funktsional. Anal. i Prilozhen. **23** (1989), no. 1, 1–23; II, **23**, no. 4, 53–62; English transl. in Functional Anal. Appl. **23** (1989).

7. V. P. Maslov, *New superposition principle for optimization problems*, Seminaire Equations aux derivées partielles, 1985–1986, Exp. XXIV, École Polytech., Palaiseau, 1986.

8. ———, *On a new superposition principle for optimization problems*, Uspekhi Mat. Nauk **42** (1987), no. 3, 39–48; English transl. in Russian Math. Surveys **42** (1987).

9. S. N. Samborskiĭ and G. B. Shpiz, *Convex sets in the semimodule of bounded functions*, this volume.

CENTER FOR OPTIMIZATION AND MATHEMATICAL MODELING, INSTITUTE OF NEW TECHNOLOGIES, KIROVOGRADSKAYA UL., 11, 113587, MOSCOW, RUSSIA

Translated by I. PETRENKO

ADVANCES IN SOVIET MATHEMATICS
Volume 13, 1992

Convex Sets in the Semimodule of Bounded Functions

S. N. SAMBORSKIĬ AND G. B. SHPIZ

In this paper the analog of a classical fact of functional analysis about characterization of convex sets is obtained; terminology from [1] is used.

Let K be an idempotent semiring with invertible multiplication such that any bounded subset belonging to it has a supremum, X a set, and $\mathscr{B}(X, K)$ the semimodule of bounded K-valued functions defined on X. For $f, g \in \mathscr{B}(X, K)$ suppose $\langle f, g\rangle = \sup_{x\in X} f(x) \odot g(x)$. Denote by e the function from $\mathscr{B}(X, K)$ which is equal to $\mathbf{1} \in K$ identically.

DEFINITION 1. A set $M \subset \mathscr{B}(X, K)$ is called *convex*, if $a\odot x\oplus b\odot y \in M$ for any $x, y \in M$ and $a, b \in K$ such that $a \oplus b = \mathbf{1}$.

The idempotence of K implies that the inclusion $\sup_i x_i \in M$ is valid for any finite collection $\{x_i \in M\}$.

DEFINITION 2. A set $M \subset \mathscr{B}(X, K)$ is called *closed*, if

a) any bounded subset in M has a supremum belonging to M;
b) any element $x \in \mathscr{B}(X, K)$ such that the segment $[x, d \odot x]$ intersects M for all $d \in K$, $d > \mathbf{1}$ belongs to M.

For the set M to be closed, it is sufficient, for example, that this set be closed in the topology of pointwise convergence of functions with respect to the interval topology in K.

DEFINITION 3. A *closed affine semispace* is defined as any subset $M \subset \mathscr{B}(X, K)$ of the form:

$$M = \{x \in \mathscr{B}(X, K) \mid \langle a, x\rangle \oplus k \geqslant \langle a', x\rangle \oplus k'\},$$

where $k, k' \in K$, $a, a' \in \mathscr{B}(X, K)$.

THEOREM 1. *A closed convex set coincides with the intersection of all closed affine semispaces that contain it.*

PROOF. Since multiplication is invertible, it is sufficient to prove that if e does not belong to a closed convex set M, then there exists an affine semispace that contains M and does not contain e.

1991 *Mathematics Subject Classification.* Primary 16Y20, 47B38.

For $f \in \mathscr{B}(X, K)$, assume that

$$N(f) = \sup_{x\in X}(f \oplus e)(x) = \mathbf{1} \oplus \sup_{x\in X} f(x) \in K,$$
$$M(d) = \{f \in M \mid N(f) \leqslant d\}, \qquad D = \{d \in K \mid M(d) \neq \varnothing\},$$
$$d_0 = \inf_{d\in D} d, \quad \widehat{F}_d = \sup_{f\in M(d)} f, \quad F_d = d^{-1} \odot \widehat{F}_d, \quad \Phi = \sup_{d\in D} F_d.$$

Evidently the set D is nonempty, $d_0 \geqslant \mathbf{1}$ and $M = \bigcup_{d\in D} M(d)$. Moreover for any $x \in X$, $F_d(x) \leqslant N(\widehat{F}_d) \leqslant d$.

Lemma 1. *The following inclusions and inequalities are valid*:

(1) $\widehat{F}_d \subset M(d)$;
(2) $F_d \leqslant F_c$ *for* $c \leqslant d$, $c, d \in D$;
(3) $\widehat{F}_d \leqslant d \odot \Phi$ *for* $d \in D$;
(4) $\widehat{F}_d \geqslant \Phi$ *for* $d \in D$.

Proof. (1) follows from the fact that M is closed.

(2) By virtue of the convexity of M, we have: $c \odot d^{-1} \odot \widehat{F}_d \oplus \widehat{F}_c \in M$ and $N(c \odot d^{-1} \odot \widehat{F}_d \oplus \widehat{F}_c) \leqslant c$, whence $(c \odot d^{-1} \odot \widehat{F}_d \oplus \widehat{F}_c) \in M(c)$, and therefore, $(c \odot d^{-1} \odot \widehat{F}_d \oplus \widehat{F}_c) \leqslant \widehat{F}_c$. The last relation yields the required inequality.

(3) $\Phi \geqslant d^{-1} \odot \widehat{F}_d$ by construction.

(4) It is sufficient to verify that $\widehat{F}_d \geqslant F_c = c^{-1} \odot \widehat{F}_c$ for $c, d \in D$. But for any $x \in X$ we have

$$c^{-1} \odot \widehat{F}_c(x) \oplus \widehat{F}_d(x) \oplus e \leqslant c^{-1} \odot N(\widehat{F}_c) \oplus N(\widehat{F}_d) \leqslant d,$$

whence $N(c^{-1} \odot \widehat{F}_c \oplus \widehat{F}_d) \leqslant d$.

On the other hand, $(c^{-1} \odot \widehat{F}_c \oplus \widehat{F}_d) \in M$ by virtue of the convexity of M, whence $(c^{-1} \odot \widehat{F}_c \oplus \widehat{F}_d) \in M(d)$.

Thus we obtain the required inequality.

Lemma 2. *If* $M(\mathbf{1})$ *is empty, then there exists a* $d > \mathbf{1}$ *such that* $N(f) \geqslant d$ *for all* $f \in M$.

Proof. If $d_0 = \mathbf{1}$, then from Lemma 1 and the fact that M is closed we get $\Phi \in M$. But this is impossible, since $N(\Phi) \leqslant \mathbf{1}$ by construction. Therefore $d_0 > \mathbf{1}$. Thus the proof is complete.

By virtue of Lemma 1, to prove the theorem it is sufficient to consider the case when $M(\mathbf{1})$ is nonempty.

Since $e \notin M$, we have $F_1(x_0) < \mathbf{1}$ for some $x_0 \in X$. Define the function $a \in \mathscr{B}(X, K)$ by the formula

$$a(x) = \begin{cases} F_1(x)^{-1} & \text{if } x = x_0, \\ 0 & \text{otherwise.} \end{cases}$$

LEMMA 3. *For any* $f \in M$ *the inequality* $\langle a, f\rangle \leqslant N(f)$ *is satisfied.*

PROOF. Indeed, since $N(f) \geqslant \mathbf{1}$, we have $N(f)^{-1} \oplus \mathbf{1} = \mathbf{1}$, whence by the convexity of M and Lemma 1 it follows that $N(f)^{-1} \odot f \oplus F_1 \in M(\mathbf{1})$. And so $N(f)^{-1} \odot f \leqslant F_1$. Whence $f \leqslant N(f) \odot F_1$, and therefore,

$$\langle a, f\rangle = F_1(x_0)^{-1} \odot f(x_0) \leqslant F_1(x_0)^{-1} \odot N(f) \odot F_1(x_0) = N(f),$$

as required.

Now to prove the theorem it remains to note that

$$\langle a, e\rangle = F_1(x_0)^{-1} > N(e).$$

REFERENCES

1. P. I. Dudnikov and S. N. Samborskiĭ, *Endomorphisms of finitely generated free semimodules*, this volume.

CENTER FOR OPTIMIZATION AND MATHEMATICAL MODELING, INSTITUTE OF NEW TECHNOLOGIES, KIROVOGRADSKAYA UL., 11, 113587, MOSCOW, RUSSIA

Translated by I. PETRENKO

ADVANCES IN SOVIET MATHEMATICS
Volume 13, 1992

The Fourier Transform and Semirings of Pareto Sets

S. N. SAMBORSKIĬ AND A. A. TARASHCHAN

Numerous optimization problems lead to the investigation of endomorphisms acting in functional semimodules over idempotent semirings (i.e., endomorphisms acting in Maslov spaces). If the semiring is not numerical or, in a more general case, does not satisfy the stabilization condition that was introduced in [2], then the spectral analysis of such endomorphisms can be too complicated. Nevertheless, a number of consequences of spectral theory (for example, estimates for iterates of endomorphisms) can be obtained by applying the "Fourier transform", which is the complete analog of the classical Fourier transform but is defined for idempotent operations.

In the papers [3, 4] V. P. Maslov remarked that the Fourier idempotent integral over the family of "eigenfunctions" of the translation operator (i.e., of the idempotent analog of exponents; linear functions are the same) coincides with the Legendre transform.

In this paper we introduce an abstract Fourier transform for an important class of semirings. In some cases this transform becomes the classical Legendre transform. Semirings of Pareto sets belong to this class. Using the Fourier-Legendre transform, we find two-sided estimates for the dynamics of Pareto sets for multicriterial optimization problems.

Let G be a closed subgroup of the group $\mathbb{R}^n$ with respect to addition and K be a semiring with an idempotent operation $\oplus$ [2]. Recall that the operation $\oplus$ induces a partial order in the semiring K according to the rule:

$$a \leqslant b \quad \text{if } a \oplus b = b.$$

DEFINITION 1. The semiring $P(G, K)$ is defined as the set of bounded (in the sense of the order in K) mappings from G to K with the operations

$$\begin{aligned} (\varphi \oplus \Psi)(g) &= \varphi(g) \oplus \Psi(g), \qquad g \in G, \\ (\varphi \odot \Psi)(g) &= \bigoplus \int_G \Psi(g-h) \odot \varphi(h)\, dh = \sup_{h \in G}(\Psi(g-h) \odot \varphi(h)). \end{aligned} \tag{1}$$

1991 *Mathematics Subject Classification*. Primary 16Y20, 47B38; Secondary 49B99.

Let e be a mapping from G to K such that

$$\bigoplus\int_G e(g-h)\odot e(h)dh = e(g)$$

(i.e., e is an idempotent element with respect to the multiplication in the semiring $P(G, K)$).

DEFINITION 2. The semiring $P_E(G, K)$ is defined as the set of mappings from G to K of the form

$$\bigoplus\int_G e(g-h)\odot \Psi(h)\,dh$$

for all possible $\Psi \in P(G, K)$ with the operations given by (1).

The element e is the unit $\mathbf{1}$ in the semiring $P_e(G, K)$.

The semiring $P(G, K)$ may be regarded as the semiring $P_e(G, K)$ if we take e to be the δ-function: $\delta_0(x) = \mathbf{0}$ for $x \neq 0 \in G$ and $\delta_0(0) = \mathbf{1}$.

EXAMPLE 1. The semirings $P_e(G, K)$ (in particular, the semiring from Definition 1) arise in problems connected with the synthesis of discrete computational media. We give a brief description of some objects connected with such media; the details can be found in the monograph [1].

Let X be some finite set, $\mathbb{Z}^3$ the integer lattice in $\mathbb{R}^3$, $\Omega = X \times \mathbb{Z}^3$, K a semiring, and $L\colon \Omega \times \Omega \to K$ a mapping (the *characteristic of connections* within the medium) that satisfies the following conditions (of homogeneity, localness, and regularity): there exists a mapping $L_0\colon \mathbb{Z}^3 \times X \times X \to K$ (the *characteristic* of the medium) such that

$$L((x, g), (y, h)) = L_0(g-h, x, y),$$

and the support V (the *regulator* of the medium) of the following mapping: $L_0(\cdot, x, y)\colon \mathbb{Z}^3 \to K$ is independent of x and y and is bounded.

The semimodule $\mathscr{B}(X, K)$ is the state space of the medium. The *architecture* (X, V) of the medium and its characteristic L_0 determine the endomorphism

$$(\mathscr{L}v)(x, g) = \bigoplus_{h\in\mathbb{Z}^3}\bigoplus_{y\in X}(L_0(g-h, x, y)\odot v(y, h))$$

on the state space $\mathscr{B}(X, K)$ of the medium.

PROPOSITION 1. *In the semimodule $\mathscr{B}(X, P(\mathbb{Z}^3, K))$ over the semiring $P(\mathbb{Z}^3, K)$, the endomorphism $\mathscr{L}$ induces an endomorphism by the rule*

$$(Av)(x) = (\mathscr{L}v)(x, \cdot), \qquad v \in \mathscr{B}(X, P(\mathbb{Z}^3, K)),$$

that is an integral operator with kernel in the space $\mathscr{B}(X \times X, P(\mathbb{Z}^3, K))$.

The proof immediately follows from the special form of the endomorphism $\mathscr{L}$.

REMARK 1. This example illustrates one of the attractive features of idempotent analysis, namely: the set of semiring-valued functions has a semiring structure itself, and the operators acting on functions, say, of two variables, may be regarded as operators acting on functions of one variable with values in the functional semiring consisting of functions of the other variable. Certainly, this method is one of the most popular in applications to ordinary functional analysis. There may not be many fields, but there is a lot of semirings, and we may often apply general theorems of idempotent analysis to new objects, since such an approach preserves the semimodule over semiring structure. However, the algebraic properties of functional semirings are, clearly, much worse than, for example, those of numerical semirings or semirings with a finite number of generators.

Other examples of semirings of the form $P_e(G, K)$ are connected with multicriterial optimization problems.

Let the semiring K be isomorphic to a closed subsemiring of the semiring $\mathscr{R}$ (see Example 2.1 in the paper [2]). We shall assume that $K \subseteq \mathscr{R}$. Let $K_+ = K \setminus \{\mathbf{0}\}$ and $\bar{K}^m = (K_+)^m \cup \{(\mathbf{0}, \mathbf{0}, \dots, \mathbf{0})\}$. Suppose further that there is a partial order $\leqslant$ in $\bar{K}^m$ compatible with addition in $\bar{K}^m$ and with multiplication by $\lambda \in K$ in the following sense:

(1) $\alpha \leqslant \beta \Rightarrow \alpha \oplus \gamma \leqslant \beta \oplus \gamma$ for any $\alpha, \beta, \gamma \in \bar{K}^m$;
(2) $\alpha \leqslant \beta \Rightarrow \lambda \odot \alpha \leqslant \lambda \odot \beta$ for any $\alpha, \beta \in \bar{K}^m$ and $\lambda \in K$;
(3) $\lambda \leqslant \mu$ (in the semiring K) $\Rightarrow \lambda \odot \alpha \leqslant \mu \odot \alpha$ for any $\lambda, \mu \in K$ and $\alpha \in \bar{K}^m$.

DEFINITION 3. If A is a subset of $\bar{K}^m$, then $\max A$ denotes the set of all maximal elements of the closure of A in the topology in $\bar{K}^m$ induced from $\mathbb{R}^m$ (with respect to the partial order $\leqslant$).

DEFINITION 4. The semiring $P(K^m, \leqslant)$ is defined as the set of bounded subsets A in $\bar{K}^m$ such that $\max A = A$ with the operations

$$A \oplus B = \max(A \cup B) \quad \text{and} \quad A \odot B = \max(A + B),$$

where

$$A + B = \{a + b \mid a \in A, b \in B\}.$$

EXAMPLE 2. Let $n > 1$ be a positive integer. There is a partial order $\leqslant$ in the space $\mathbb{R}^n$:

$$x = (x_1, x_2, \dots, x_n) \leqslant y = (y_1, y_2, \dots, y_n) \iff x_i \leqslant y_i \ (\forall i = 1, \dots, n);$$

$\max A$ is defined as the operation which assigns to every subset A of $\mathbb{R}^n$ the set of maximal elements of the closure of A.

Now consider the set $P(\mathscr{R}_+^n)$ of bounded subsets A of $\mathbb{R}^n \cup \{-\infty\}$ such that $\max A = A$ and define the operations

$$A \oplus B = \max(A \cup B), \qquad A \odot B = \max(A + B),$$

where

$$A + B = \{x \in \mathbb{R}^n \mid x = a + b,\ a \in A,\ b \in B\}.$$

The set $P(\mathscr{R}_+^n)$ with these operations becomes a commutative semiring in which $\mathbf{0}$ coincides with the one-point set $\{-\infty\}$ and $\mathbf{1}$ coincides with the one-point set $\{0\}$.

Standard multicriterial problems of dynamic programming are also associated with an arbitrary order $\leqslant$ admissible in the sense indicated above.

Suppose that a controlled dynamic process is given by the mapping f: $X \times U \to X$. The choice of $x_0 \in X$ and the collection of controls $\{u_1, \dots, u_k\}$ ($u_i \in U$) determine in X an admissible trajectory $\pi_k = (x_0, \dots, x_k)$, where $x_i = f(x_{i-1}, u_i)$. Let φ be a mapping belonging to $\mathscr{B}(X \times U, \bar{K}^m)$ that determines a "quality criterion" on the trajectories π_k by the rule

$$\Phi_k(\pi_k) = \varphi(x_0, u_1) \odot \varphi(x_1, u_2) \odot \cdots \odot \varphi(x_{k-1}, u_k).$$

DEFINITION 5. The *Pareto set* $\omega_k(x)$ for the criterion Φ on the set Θ of admissible trajectories is defined as the subset

$$\max\left(\bigcup_{\pi_k} \Phi(\pi_k)\right),$$

where the π_k run through all possible k-step admissible trajectories (starting at the point x) in the set Θ.

The optimality principle immediately implies a recursion relation for the Pareto sets for the criterion Φ:

$$\omega_{k+1}(x) = \max\left(\bigcup_{u \in U} (\varphi(x, u) \odot \omega_k(f(x, u)))\right).$$

In the semimodule $\mathscr{B}(X, P(K^m, \leqslant))$ over the semiring $P(K^m, \leqslant)$ determine the mapping A by the rule:

$$(A\omega)(x) = \max\left(\bigcup_{u \in U} (\varphi(x, u) \odot \omega(f(x, u)))\right). \tag{2}$$

It is obvious that if $X' \subseteq X$ and $\omega_0(x) = \mathbf{1}$ for $x \in X'$ and $\omega_0(x) = \mathbf{0}$ for $x \notin X'$, then $(A^k \omega_0)(x)$ is the Pareto set for the criterion Φ on the set of trajectories beginning in X'.

PROPOSITION 2. *The mapping A given by formula* (2) *is an endomorphism of the semimodule $\mathscr{B}(X, P(K^m, \leqslant))$ over the semiring $P(K^m, \leqslant)$.*

The proof is straightforward.

PROPOSITION 3. *The semiring $P(K^m, \leqslant)$ (see Definition* 4) *is isomorphic to the semiring $P_e(G, K)$ for some $e\colon G \to K$ depending on the partial order $\leqslant$, where G is isomorphic to the group K_+^{m-1}.*

PROOF. The group K_+^m can be regarded as the direct sum of the subgroup $\{\lambda \odot (\mathbf{1}, \mathbf{1}, \dots, \mathbf{1}) \mid \lambda \in K\}$ and a subgroup G isomorphic to K_+^{m-1}. For every Pareto set C and every $g \in G$, the set

$$V_{c,g} = \{\lambda \in K \mid \exists c \in C\colon \lambda \odot g \leqslant c\}$$

has the following form:

$$V_{c,g} = \{\lambda \in K \mid \lambda \leqslant \lambda_{c,g} \in K\};$$

this immediately follows from the definition of Pareto set and the compatibility of the partial order $\leqslant$ with the operations in $\bar{K}^m$. And so for every C we have the mapping $g \to \lambda_{c,g}$ which is an element of the semiring $P(G, K)$. In particular, taking the one-point Pareto set $\{(\mathbf{1}, \mathbf{1}, \dots, \mathbf{1})\}$ ($(\mathbf{1}, \mathbf{1}, \dots, \mathbf{1}) = \mathrm{E} \in \bar{K}^m$) we obtain the mapping $e\colon g \to \lambda_{\mathrm{E},g}$. This mapping e is idempotent with respect to the multiplication in $P(G, K)$.

It can be verified directly that the semiring $P_e(G, K)$ is isomorphic to the semiring $P(K^m, \leqslant)$ for the mapping e chosen in this way.

EXAMPLE 3. Suppose $n = 2$, $K = \mathscr{R}$ and the partial order $\leqslant$ coincides with the usual pointwise partial order in $\mathbb{R}^2$ (see Example 2). We shall take e to be the mapping $x \to -|x|$. Then the semiring $P(\mathscr{R}^2)$ is isomorphic to the semiring of $\mathscr{R}$-valued functions on $\mathbb{R}^1$ of Lipschitz class with constant 1 satisfying $\sup_g(\Psi(g) + |g|) < \infty$. The isomorphism indicated in Proposition 3 can obviously be defined as follows (Figures 1a–1c): for a set α we first construct the set $\bar{\alpha} = \{a \in \mathbb{R}^2 \mid \exists b \in \alpha : a \leqslant b\}$, then rotate it in the plane $\mathbb{R}^2$ through the angle $\pi/4$ and take the boundary of the rotated set. We get a Lipschitz function.

For $G = \mathbb{Z}^3$, if we take e to be the δ-function ($\delta(0) = \mathbf{1}$, $\delta(g) = \mathbf{0}$ for $g \neq 0$), then we obtain $P_e(\mathbb{Z}^3, K) = P(\mathbb{Z}^3, K)$.

Thus, to study the endomorphisms from Propositions 1 and 2, we shall consider endomorphisms of semimodules over the semiring $P_e(G, K)$. For simplicity it will be assumed that $K = \mathscr{R}$. Let $\operatorname{Char} G$ be the set of characters of the group G, and let e be an idempotent with respect to convolution (as described above). We define Ω_e to be the subset of $\operatorname{Char} G$ consisting of elements χ such that $\sup_{g \in G}(\chi(g) + e(g)) < \infty$. For the above choice of the group $G \subset \mathbb{R}^n$ and the semiring $K = \mathscr{R}$, the set $\operatorname{Char} G$ can be monomorphically imbedded in the set of linear bounded mappings from $\mathbb{R}^n$ into $\mathbb{R}^1$; thus, $\operatorname{Char} G$ is a topological vector space with the topology induced from the adjoint space $(\mathbb{R}^n)^*$.

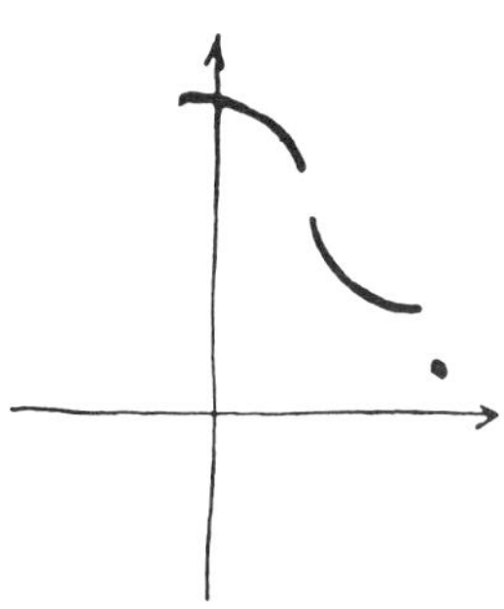

FIGURE 1a

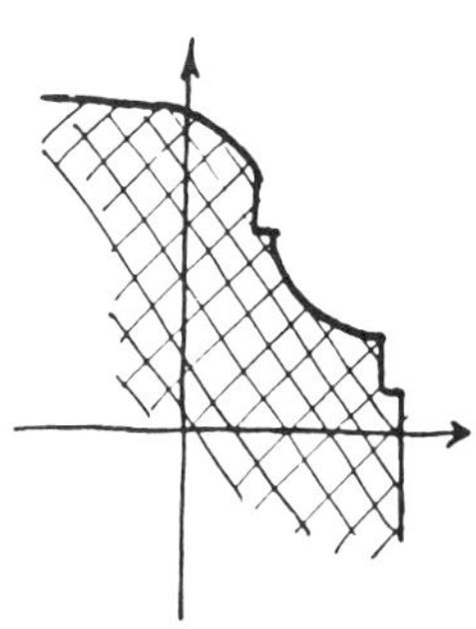

FIGURE 1b

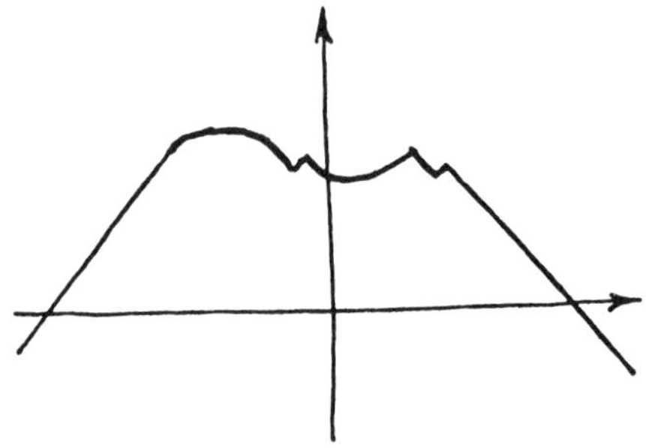

FIGURE 1c

ASSERTION 1. *Ω_e is a closed convex subset containing the zero character in the topological vector space* Char G *over* $\mathbb{R}^1$.

PROOF. Suppose λ is a real number between 0 and 1, while χ_i ($i = 1, 2$) are two characters from Ω_e. Then we have

$$\begin{aligned}
&\sup_{g\in G}(\lambda\chi_1(g) + (1-\lambda)\chi_2(g) + e(g)) \\
&\qquad = \sup_{g\in G}(\lambda(\chi_1(g) + e(g)) + (1-\lambda)(\chi_2(g)e(g))) \\
&\qquad \leqslant \lambda \sup_{g\in G}(\chi_1(g) + e(g)) + (1-\lambda)\sup_{g\in G}(\chi_2(g) + e(g)) < \infty.
\end{aligned}$$

The rest of the assertion is evident.

DEFINITION 6. The *Fourier transform* $F(a)$ for each function a from the semiring $P_e(G, \mathscr{R})$ is defined to be the $\mathscr{R}$-valued function

$$(Fa)(\chi) = \bigoplus\int_G \chi(g)\odot a(g)\,dg = \sup_{g\in G}(\chi(g) + a(g))$$

on Ω_e.

In the case of Example 3, the Fourier transform coincides with the Legendre transform.

EXAMPLE 4. Let e be an idempotent element with respect to convolution ($e * e = e$) in the set of $\mathscr{R}$-valued functions on a group $G \subset \mathbb{R}^n$ and Ω_e be defined as described above. Denote by $\mathrm{Conc}(\Omega_e)$ the semiring of concave $\mathscr{R}$-valued functions on Ω_e with the operations of pointwise maximum ($\oplus$) and pointwise addition ($\odot$):

$$(f\oplus g)(x) = f(x)\oplus g(x), \qquad (f\odot g)(x) = f(x)\odot g(x).$$

The functions identically equal to $\mathbf{0}\in\mathscr{R}$ and $\mathbf{1}\in\mathscr{R}$ are, respectively, the neutral elements in $\mathrm{Conc}(\Omega_e)$.

PROPOSITION 4. *The Fourier transform is a homomorphism from the semiring $P_e(G, \mathscr{R})$ into the semiring* $\mathrm{Conc}(\Omega_e)$ *with the following properties*:

a) *F is an epimorphism*;

b) *the mapping given by the formula*

$$(\mathcal{L} f)(g) = \inf_{\chi \in \Omega_e} (-\chi(g) + f(\chi)),$$

is a semiring monomorphism of $\mathrm{Conc}(\Omega_e)$ *into the* $P_e(G, \mathcal{R})$ *and the one-sided inverse of* F *; moreover, the range of* $\mathcal{L}$ *coincides with the set of functions in* $P_e(G, \mathcal{R})$ *convex on* G *;*

c) *in the inverse image* $F^{-1}(a)$ *of each element* $a \in \mathrm{Conc}(\Omega_e)$ *as a subset of the partially ordered set* $P_e(G, \mathcal{R})$ *, there exist a maximal element* $\mathcal{L}(a)$ *and a minimal element* $\mathcal{H}(a)$ *.*

PROOF. Let $a, b \in P_e(G, \mathcal{R})$. Then

$$\begin{aligned} F(a \oplus b)(\chi) &= \sup_{g \in G}(\chi(g) + \max(a(g), b(g))) \\ &= \max\Big(\sup_{g \in G}(\chi(g) + a(g)), \sup_{g \in G}(\chi(g) + b(g))\Big) \\ &= (F(a) \oplus F(b))(\chi), \\ F(a * b)(\chi) &= \sup_{g \in G}(\chi(g) + (a * b)(g)) \\ &= \sup_{g \in G}\Big(\chi(h) + \chi(g-h) + \sup_{g \in G}(a(h) + b(g-h))\Big) \\ &= \sup_{g \in G}(\chi(h) + a(h)) + \sup_{g \in G}\sup_{h \in G}(\chi(g-h) + b(g-h)) \\ &= F(a)(\chi) \odot F(b)(\chi), \end{aligned}$$

i.e., F is a homomorphism (from the semiring $P_e(G, \mathcal{R})$ into the semiring $\mathrm{Conc}(\Omega_e)$).

The image $F(a)$ of each element $a \in P_e(G, \mathcal{R})$ is a concave function. Indeed, for some real number λ such that $0 \leqslant \lambda \leqslant 1$ and $\chi_1, \chi_2 \in \Omega_e$ we have

$$\begin{aligned} &\lambda F(a)(\chi_1) + (1-\lambda) F(a)(\chi_2) \\ &\quad = \sup_{g \in G}(\lambda \chi_1(g) + \lambda a(g)) + \sup_{g \in G}((1-\lambda)\chi_2(g) + (1-\lambda) a(g)) \\ &\quad \geqslant \sup_{g \in G}(\lambda \chi_1(g) + (1-\lambda)\chi_2(g) + a(g)) = F(a)(\lambda \chi_1 + (1-\lambda)\chi_2). \end{aligned}$$

Assertion a) follows from Assertion c). To prove Assertion c), we shall fix a function f that is concave on Ω_e. Then

$$\begin{aligned} F(\mathcal{L} f)(\chi) &= \sup_{g \in G}\Big(\chi(g) + \inf_{\Theta \in \Omega_e}(-\Theta(g) + f(\Theta))\Big) \\ &= \inf_{\Theta \in \Omega_e}\Big(\sup_{g \in G}(\chi - \Theta)(g) + f(\Theta)\Big). \end{aligned}$$

It is easy to see that the infimum on the right-hand side of the last expression is achieved for $\Theta = \chi$ and equals $f(\chi)$.

The fact that $\mathscr{L}(f)$ is a convex function can be proved similarly to the fact that $F(a)$ is a concave function.

Finally, consider the inverse image $F^{-1}(f)$ of an element f from the set $\operatorname{Conc}(\Omega_e)$. Since the function $a = \mathscr{L}(f)$ is convex, we see that for every function $\bar{a}$ defined on G such that at some point $\bar{x} \in G$ we have the inequality $\bar{a}(\bar{x}) > a(\bar{x})$, the image $F(\bar{a})$ does not coincide with the image $F(a)$, i.e., a is a maximal element in $F^{-1}(f)$. Consider the undergraph of this function a, i.e., the set

$$\alpha = \{(g, z) \in G \times \mathbb{R}^1 \mid z \leqslant a(g)\},$$

which is convex. Let A be a minimal (with respect to inclusion) set of points from G such that α is the convex hull in $G \times \mathbb{R}^1$ of the set

$$\mathscr{M} = \bigcup_{h \in A} U_h,$$

where U_h is the undergraph of the function $g \to e(g-h) \odot a(h)$ that acts from G into $\mathbb{R}^1$. Then A is uniquely determined and the boundary of the set $\mathscr{M}$ is the graph of the function $b\colon G \to \mathscr{R}^1$ which belongs (by construction) to the space $P_e(G, \mathscr{R}^1)$. It is easy to see that b is a minimal element in $F^{-1}(f)$. Therefore, the assertion is proved.

Let A be an endomorphism of the semimodule $\mathscr{B}(X, P_e(G, \mathscr{R}))$ over the semiring $P_e(G, \mathscr{R})$. To this endomorphism we assign the following collection $\{A_\chi\colon \mathscr{B}(X, \mathscr{R}) \to \mathscr{B}(X, \mathscr{R}) \mid \chi \in \Omega_e\}$ of endomorphisms acting in $\mathscr{B}(X, \mathscr{R})$. In the case when X is finite and A is written as a matrix (a_{ij}) with elements in $P_e(G, \mathscr{R})$, A_χ can be written as the matrix $((Fa_{ij})(\chi))$, where F is the Fourier transform. In the case when X is arbitrary, we define the homomorphism Ψ from the semimodule $\mathscr{B}(X, P_e(G, \mathscr{R}))$ over the semiring $P_e(G, \mathscr{R})$ into the semimodule $\mathscr{B}(X \times \Omega_e, \mathscr{R})$ over the semiring $\mathscr{R}$ in a natural way. Let a be a function belonging to the semimodule $\mathscr{B}(X, P_e(G, \mathscr{R}))$. For a fixed x the element $a(x)$ is an element of $P_e(G, \mathscr{R})$, and $F(a(x))(\cdot)$ (i.e., when the Fourier transform is applied) is a function of $\chi \in \Omega_e$. For a fixed $\chi \in \Omega_e$, we get an element $F(a(x))(\chi)$ that belongs to $\mathscr{R}$. Thereby for every function $a \in \mathscr{B}(X, P_e(G, \mathscr{R}))$ a function assigning to the pair (x, χ) the element $F(a(x))(\chi)$ of $\mathscr{R}$ is defined and this function, which will be denoted in the sequel by Ψa, belongs to the space $\mathscr{B}(X \times \Omega_e, \mathscr{R})$.

For an endomorphism A on the semimodule $\mathscr{B}(X, P_e(G, \mathscr{R}))$ there exists a unique endomorphism $F(A)$ such that the following commutation relation $\Psi A = F(A)\Psi$ holds. When fixing $\chi \in \Omega$, the set of functions $v(\cdot, \chi)$ from the range of the homomorphism χ coincides with the semimodule $\mathscr{B}(X, \mathscr{R})$.

Now suppose that the operators belonging to the collection

$$\{A_\chi\colon \mathscr{B}(X, \mathscr{R}) \to \mathscr{B}(X, \mathscr{R}) \mid \chi \in \Omega_e\}$$

act as follows:

$$A_\chi v(x, \chi) = (F(A)v)(x, \chi).$$

It is easy to see that by the definition of the mappings Ψ and $F(A)$ this assumption is correct.

According to the construction from §7 of the paper [2] to each endomorphism A_χ for a fixed $\chi \in \Omega_e$ there correspond:

a) the decomposition $X = \bigcup X_i$ of the set X into nonintersecting subsets X_i;

b) the spectral function $\lambda_Y(\cdot, \chi)$ corresponding to a subset $Y \subset X$.

The spectral function permits to obtain two-sided estimates of iterates of an endomorphism A_χ as described in Theorem 7.1 of the paper [2].

Now we want to establish that the spectral functions depend on χ as a parameter.

PROPOSITION 5. *Assume that X is a finite set, A is an endomorphism of the semimodule $\mathscr{B}(X, P_e(G, \mathscr{R}))$, and $\{A_\chi : \mathscr{B}(X, \mathscr{R}) \to \mathscr{B}(X, \mathscr{R}) \mid \chi \in \Omega_e\}$ is the collection of endomorphisms (acting in the semimodule $\mathscr{B}(X, \mathscr{R})$) that was constructed above. Then*:

a) *the decomposition $X = \bigcup X_i$ corresponding to the endomorphism A_χ can be taken to be the same for all $\chi \in \Omega_e$;*

b) *the spectral functions $\lambda_Y(x, \chi)$ of the endomorphisms A_χ for each fixed x are concave on Ω_e as functions of χ.*

PROOF. Note, firstly, that if some function a belonging to the semimodule $P_e(G, \mathscr{R})$ is not equal to $\mathbf{0}$ identically, then for any $\chi \in \Omega_e$ the value $F(a)(\chi)$ of its Fourier transform at the point χ is not equal to $\mathbf{0}$. Indeed, from the definition of the $\mathscr{R}$-valued character of the group G, it follows that for any $\chi \in \Omega_e$ which is not $\mathbf{0}$ identically and for any $g \in G$, the inequality $\chi(g) \neq 0$ is valid; otherwise for any $g_1 \in G$ the following series of inequalities

$$\chi(g_1) = \chi(g_1 - g + g) = \chi(g_1 - g) \odot \chi(g) = \mathbf{0}$$

would hold. So, if for some $\chi \in \Omega_e$ the inequality $a_{ij}(\chi) \neq \mathbf{0}$ is valid, then this inequality is valid also for any $\chi \in \Omega_e$. By the construction of the decomposition $X = \bigcup X_i$ for the endomorphism $\alpha = (\alpha_{ij})$, the values α_{ij} themselves are not important, but only the fact whether these values are $\mathbf{0}$ or not is essential [2]. And so the decompositions $X = \bigcup X_i$ corresponding to the endomorphisms A_χ with the matrices

$$\{F(a_{ij})(\chi) \mid i, j = 1, \dots, n\},$$

where $\{a_{ij} \mid i, j = 1, \dots, n\}$ is the matrix of the endomorphism A over the semiring $P_e(G, \mathscr{R})$, do not depend on χ.

The operation $\oplus$ in the semiring $\mathscr{R}$ is such that for concave functions f_1 and f_2 defined on Ω_e their pointwise sum $f_1 \oplus f_2$ is also a concave function. By the definition of the spectral function $x \to \lambda_Y(x)$ of the endomorphism

α from [2], for fixed x and Y the values of this function are sums (in the semiring $\mathscr{R}$) of the eigenvalues of the endomorphism α. And so to prove the concavity with respect to χ of the spectral function $x \to \lambda_Y(x, \chi)$ of the endomorphism A_χ, it is sufficient to consider the particular case of an indecomposable endomorphism A_χ for some (and in virtue of that was stated above, for every) $\chi \in \Omega_e$. In this case A_χ possesses a unique eigenelement λ_χ, and Proposition 5 will be proved if we show that λ_χ is a concave function of $\chi \in \Omega_e$. According to the formula for the eigenelement λ of the endomorphism α [2], i.e., $\lambda = \bigoplus \lambda_{(k)}$, where $\lambda^k_{(k)} = \operatorname{Tr} \alpha^k = \bigoplus \alpha_{ii_1} \odot \alpha_{i_1 i_2} \odot \cdots \odot \alpha_{i_k i}$, the element λ_χ can be obtained from the elements $F(a_{ij})(\chi)$ of the matrix A_χ by means of the operations $\odot$ and $\oplus$ that preserve the concavity of functions to which they are applied pointwise. Since the functions $F(a_{ij})(\cdot)$ are concave with respect to χ, we see that the mapping $\chi \to \lambda_\chi$ is also concave and the proposition is proved.

Proposition 5 allows us to give the following definition.

DEFINITION 7. The *spectral function of an endomorphism A on the semimodule $\mathscr{B}(X, P_e(G, \mathscr{R}))$ corresponding to a subset $Y \subset X$* (X is finite) is defined as the mapping $\Lambda_Y : X \to P_e(G, \mathscr{R})$ given by the following rule:

$$\Lambda_Y(x) = \mathscr{L}(\lambda_Y(x, \cdot)),$$

where $\lambda_Y(\cdot, x)$ is the spectral function of the endomorphism A_χ acting in the semimodule $\mathscr{B}(X, \mathscr{R})$ and $\mathscr{L}$ is the inverse Fourier transform that exists due to property b) from Proposition 5.

For each element $f \in P_e(G, \mathscr{R})$ denote by $\mathscr{M}(f)$ the minimal element $f' \in P_e(G, \mathscr{R})$ such that $F(f) = F(f')$. From Proposition 4 it follows that $\mathscr{M}(f) = \mathscr{H}(Ff)$ and the mapping $\mathscr{M}$ is well defined.

THEOREM 1. *Let A be an endomorphism of the semimodule $\mathscr{B}(X, P_e(G, \mathscr{R}))$ over the semiring $P_e(G, \mathscr{R})$, X be finite, and the function $v \in \mathscr{B}(X, P_e(G, \mathscr{R}))$ satisfy the relation*

$$\operatorname{Supp}(A^s v) = \operatorname{Supp}(A^{s+1} v)$$

for some positive integer s, where $\operatorname{Supp} f = \{x \in X \mid f(x) \neq \mathbf{0} \in P_e(G, \mathscr{R})\}$. Then there exist $a, b \in P_e(G, \mathscr{R})$ with $b > \mathbf{0}$ and a positive integer N such that for all $x \in X$ and all positive integers k the estimate

$$\mathscr{M}(b \odot (\Lambda^k_Y(x))) \leqslant (A^{N+k} v)(x) \leqslant a \odot \Lambda^k_Y(x)$$

holds, where $\Lambda_Y(\cdot)$ is the spectral function of the endomorphism A corresponding to the subset $Y = \operatorname{Supp}(A^s v)$, $\leqslant$ is the partial order in the semiring $P_e(G, \mathscr{R})$, and the mapping $\mathscr{M}$ is as defined above.

PROOF. For any $\chi \in \Omega_e$, by the assumption of the theorem, the equality

$$\operatorname{Supp} A^s_\chi F(v)(\chi) = \operatorname{Supp} A^{s+1}_\chi F(v)(\chi)$$

holds, since $(A^s v)(x) \neq \mathbf{0}$ for $x \in X$ if and only if $F(A^s v(x))(\chi) \neq \mathbf{0}$ for all $\chi \in \Omega_e$. And so for any fixed $\chi \in \Omega_e$ one can use Theorem 7.1 from the paper [2]. Thus we conclude that there exist a positive integer $N(\chi)$ and $\alpha'(\chi), \beta'(\chi) \in \mathscr{R}$ with $\beta'(\chi) \neq \mathbf{0}$ such that the following estimate holds:

$$\beta'(\chi) \odot \lambda_Y^k(x, \chi) \leqslant A_\chi^{N+k} F(v(x))(\chi) \leqslant \alpha'(\chi) \odot \lambda_Y^k(x, \chi). \tag{3}$$

The elements $a_{ij}(\chi)$ of the matrix determined by the endomorphism A_χ depend on $\chi \in \Omega_e$ continuously. And so by the construction of $\beta'(\chi)$ and $\alpha'(\chi)$ in the proof of Theorem 7.1 from the paper [2] for a compact set Ω_e one can take functions $\alpha''(\cdot)$ and $\beta''(\cdot)$ continuous on Ω_e such that $\beta''(\chi) \leqslant \alpha'(\chi)$ and $\alpha'(\chi) \leqslant \alpha''(\chi)$ with $\beta''(\chi) > \mathbf{0}$. Therefore, one can also choose functions $\alpha(\cdot)$ and $\beta(\cdot)$ belonging to $\operatorname{Conc}(\Omega_e, \mathscr{R})$, where $\beta(\chi) > \mathbf{0}$, such that inequalities (3) remain valid with $\beta(\chi)$ substituted for $\beta'(\chi)$ and $\alpha(\chi)$ for $\alpha'(\chi)$. Taking into account Proposition 4, according to which the Fourier transform is surjective and has the right inverse transform $\mathscr{L}$, one can obtain the following inequalities:

$$F(b \odot \Lambda_Y^k(x))(\chi) \leqslant F(A^{N+k} v(x))(\chi) \leqslant F(a \odot (\Lambda_Y^k(x)))(\chi), \tag{4}$$

where a and b are certain functions from $P_e(G, \mathscr{R})$, $b > \mathbf{0}$, and a is convex. Now note that the system of inequalities

$$F(f)(\chi) \leqslant F(g)(\chi) \leqslant F(h)(\chi),$$

which holds for all $\chi \in \Omega_e$ and some $f, g, h \in P_e(G, \mathscr{R})$, implies, according to Proposition 4, the inequalities

$$\mathscr{M}(f) \leqslant g \leqslant \mathscr{L}(F(h)),$$

where $\leqslant$ is the partial order in $P_e(G, \mathscr{R})$. Moreover, if the equality $h = \mathscr{L}(w)$ is satisfied for some $w \in \operatorname{Conc}(\Omega_e, \mathscr{R})$, i.e., if the function h is convex, then $\mathscr{L}(F(h)) = h$.

Applying this remark to (4), we see that the inequalities (4) are valid. The theorem is proved.

REMARK 2. The estimate from Theorem 1, in contrast to analogous estimates for iterates of endomorphisms acting in semimodules over semirings satisfying the axioms from the paper [2], contains (on the right-hand side and on the left-hand side) functions that differ from each other not only by multiplication by a constant. So to illustrate the difference between the upper and the lower estimates let us consider the semiring of Pareto sets.

Let a Pareto set α in $\mathscr{R}^2$, i.e., an element of the semiring $P(\mathscr{R}^2, \leqslant)$ have, for example, the shape presented in Figure 1a. Then the continuous curve in Figure 2a shows the Pareto set $\mathscr{L}(\mathscr{F}\alpha)$ which is the maximal element of the collection $\Theta = \{\mu \mid \mathscr{F}\mu = \mathscr{F}\alpha\}$, and the set in Figure 2b is the set $\mathscr{M}(\alpha)$, i.e., the minimal element of the set Θ. Figure 2c represents the domain to which all Pareto sets from the collection Θ belong. The estimate from Theorem 1 for every k when there are different constants on the right-hand side and on the left-hand side gives the shaded domain in Figure 2d, this domain containing the Pareto set $A^{N+k}v$.

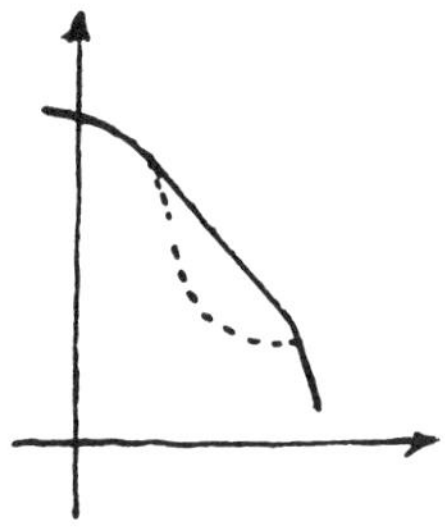

FIGURE 2a

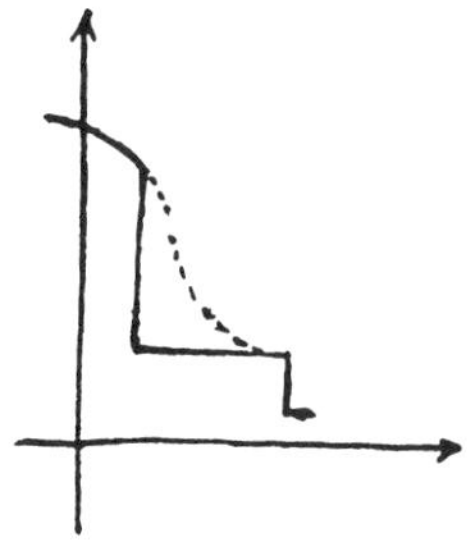

FIGURE 2b

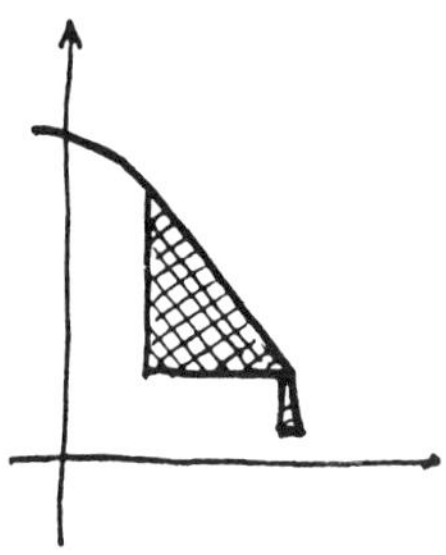

FIGURE 2c

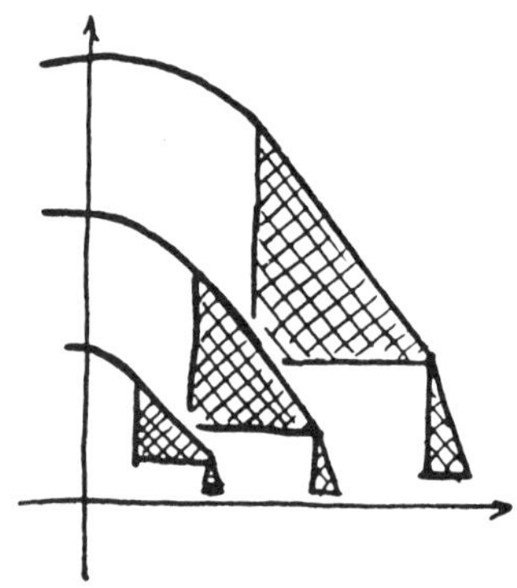

FIGURE 2d

COROLLARY 1. *Let A and v be as in Theorem* 1. *Then the equation $y = Ay \oplus v$ is solvable if and only if the spectral function $\Lambda_Y(x)$ satisfies the estimate $\Lambda_Y(x) \leqslant e$ for any $x \in X$.*

COROLLARY 2. *Suppose $\{v_i\}_{i=1}^{\infty}$ is a sequence of functions from the semimodule $\mathscr{B}(X, P_e(G, \mathscr{R}))$ that is uniformly bounded in i, i.e., $v_i \leqslant v_0$, where $\leqslant$ is the partial order in the semiring $P_e(G, \mathscr{R})$ and* $\operatorname{Supp} v_i = Y$. *Let A and Y be as in the theorem. Then the iteration process $y_{i+1} = Ay_i \oplus v_i$ is convergent if and only if the estimate $\Lambda_Y(x) \leqslant e$ holds for all $x \in X$.*

REFERENCES

1. P. I. Dudnikov and S. N. Samborskiĭ, *Endomorphisms of semimodules over semirings with an idempotent operation*, Preprint No. 87-48, Inst. Mat. Akad. Nauk Ukrain., Kiev, 1987. (Russian)

2. ———, *Endomorphisms of finitely generated free semimodules*, this volume.

3. V. P. Maslov, *Asymptotic methods for solving pseudodifferential equations*, "Nauka", Moscow, 1987. (Russian)

4. ———, *Méthodes opéracionelles*, "Mir", Moscow, 1987.

CENTER FOR OPTIMIZATION AND MATHEMATICAL MODELING, INSTITUTE OF NEW TECHNOLOGIES, KIROVOGRADSKAYA UL., 11, 113587, MOSCOW, RUSSIA

Translated by I. PETRENKO

ADVANCES IN SOVIET MATHEMATICS
Volume 13, 1992

Algebraic Remarks on Idempotent Semirings and the Kernel Theorem in Spaces of Bounded Functions

M. A. SHUBIN

Introduction

This paper contains some remarks that arose from the author's reflections about lectures given by S. N. Samborskiĭ at the seminar in the Institute of New Technologies (Moscow) in 1990.

Structure of the paper. In §1 we recall the definitions of a semiring and an idempotent semiring [1] and discuss the structure of idempotent semirings consisting of two elements (generally speaking, without neutral elements), whence, in particular, we come to the following conclusion: in the definition of an idempotent semiring the requirement that the relation $a \odot \mathbf{0} = \mathbf{0}$ be satisfied is independent of other requirements.

In §2 a classification of idempotent semirings consisting of three and four elements is given. We see that it is hardly possible to classify all finite idempotent semirings without additional restrictions, since this turns out to be too complicated. One of the restrictions that can be imposed is the cancellation condition (i.e., $a \odot c = b \odot c$, $c \neq \mathbf{0} \implies a = b$).

But, as proved in §3, the semiring consisting of two elements, namely $\mathbf{0}$ and $\mathbf{1}$, is the unique finite idempotent semiring satisfying the cancellation condition.

In §4 the concept of a free idempotent semiring with n generators is introduced and the structure of such a semiring is described. This yields, in particular, an example of a semiring in which the equality $(a \oplus b)^2 = a^2 \oplus b^2$ does not always hold, as opposed to the case of semirings satisfying the cancellation condition.

In §5 we give a description of homomorphisms of semirings with division in terms of their multiplicative kernels (the inverse images of $\mathbf{1}$), which can coincide with arbitrary convex subgroups of the multiplicative group.

1991 *Mathematics Subject Classification.* Primary 47B38.

This description leads us, in particular, to that of all homomorphisms of the semiring $\mathscr{R}_+^n = \mathbb{R}^n \cup \{\mathbf{0}\}$ and its homomorphic images (which turn out to be isomorphic to $\mathscr{R}_+^k$, where the number $k \leqslant n$ is uniquely determined).

In §6 we show that the completion method well-known in lattice theory [2] leads to the construction of an imbedding of any idempotent semiring K into an idempotent semiring $\widehat{K}$ which is a complete lattice with respect to the partial order induced by idempotent addition.

In §7 we formulate and prove the analog of the L. Schwartz theorem about kernels in spaces of bounded functions on arbitrary sets. The *normality* (monotonic continuity) introduced by P. S. Dudnikov and S. N. Samborskiĭ [1] (who realized its importance for the kernel theorem) turns out to be the necessary condition of continuity type. Note that the role of the *regularity* condition (the normality of the multiplication operator by an element of the semiring in the semiring itself) was not noticed in [1].

Note that the version of the kernel theorem (suggested here) does not involve any topology in the semiring K nor on the sets X and Y on which the functions are given. The topological version of the kernel theorem (presupposing the presence of a metric in the semiring and a topology on the sets X and Y) in spaces of continuous functions was established by V. N. Kolokoltsov and V. P. Maslov [3], [4].

Acknowledgment. I would like to express my gratitude to S. N. Samborskiĭ to whom I owe my knowledge of the principles of idempotent analysis. I am also grateful to him for numerous useful conversations.

§1. Definitions

DEFINITION 1.1. A *commutative semiring* is a set K with two associative and commutative binary operations $\oplus$ and $\odot$ satisfying the distributivity condition, i.e.,

$$a \odot (b \oplus c) = a \odot b \oplus a \odot c, \qquad a, b, c \in K, \tag{1.1}$$

possessing neutral elements $\mathbf{0}$ and $\mathbf{1}$, and satisfying the following relation:

$$a \odot \mathbf{0} = \mathbf{0}, \qquad a \in K. \tag{1.2}$$

The semiring K is called *idempotent*, if $a \oplus a = a$ for any $a \in K$.

DEFINITION 1.2. In an idempotent semiring K a *partial order* $\leqslant$ is determined as follows:

$$a \leqslant b \iff a \oplus b = b.$$

Now we omit the requirement that neutral elements exist and give a classification of idempotent semirings consisting of two elements, say, a and b. Let $a \leqslant b$. (It is clear that $a \oplus b$ equals either a or b. Since if necessary one can always substitute "a" for "b", we suppose $a \oplus b = b$, i.e., $a \leqslant b$.) Multiplying the equality $a \oplus b = b$ by a and by b, we obtain equalities that yield (by means of the partial order) the following estimate

$$a^2 \leqslant a \odot b \leqslant b^2.$$

TABLE 1

	a^2	$a \odot b$	b^2
A_1	a	a	a
A_2	a	a	b
A_3	a	b	b
A_4	b	b	b

Thus, for the multiplication $\odot$ there are four possibilities listed in Table 1.

It is easy to check that the associativity of $\odot$ and distributivity take place in all four cases. For example, due to the fact that the operations are compatible with inequalities, the distributivity $(a \oplus b) \odot c = a \odot c \oplus b \odot c$ or $b \odot c = a \odot c \oplus b \odot c$ (here $c = a$ or $c = b$) follows from the inequality

$$a \odot c \leqslant b \odot c.$$

REMARK 1.1. Evidently $a = \mathbf{0}$ and examples of the semirings A_3 and A_4 show requirement (1.2) to be independent of the rest ones. One can add $\mathbf{1}$ as a maximal element to A_3 or A_4 that proves that requirement (1.2) is independent of all other axioms for semirings with $\mathbf{1}$ as well.

§2. A classification of idempotent semirings consisting of three or four elements

THEOREM 2.1. *There exist three pairwise nonisomorphic semirings which consist of three elements (including* $\mathbf{0}$ *and* $\mathbf{1}$*). Denote by* a *the unique element that differs from* $\mathbf{0}$ *and* $\mathbf{1}$*. There exist two semirings in which* $a < \mathbf{1}$ *and the multiplication* $\odot$ *is determined by the following conditions*:

$$A_1^{(3)}: \quad a^2 = a, \qquad A_2^{(3)}: \quad a^2 = \mathbf{0}.$$

And there exists one semiring (denoted by $A_3^{(3)}$*) in which* $a > \mathbf{1}$ *and* $a^2 = a$*.*

THEOREM 2.2. *There exist* 14 *pairwise nonisomorphic idempotent semirings consisting of four elements, seven of them having* $\mathbf{1}$ *as a maximal element. List these latter ones*:

a) *there is one semiring (denoted by* $A_1^{(4)}$*) in which the elements* a *and* b *that differ from* $\mathbf{0}$ *and* $\mathbf{1}$ *are incomparable. In this semiring the operations are determined as follows*:

$$a \oplus b = \mathbf{1}, \quad a^2 = a, \quad b^2 = b, \quad a \odot b = \mathbf{0}$$

(this is the algebra of all subsets of a set consisting of two elements);

b) *there are six semirings in which* $\mathbf{0} < a < b < \mathbf{1}$ *and the multiplication* $\odot$ *is determined as in Table* 2*, where* $B_6^{(4)}$ *is the algebra of imbedded sets with the operations of union and intersection.*

If $\mathbf{1}$ *is not a maximal element, then denoting the maximal element by* b *and the other one (which differs from* $\mathbf{0}$ *and* $\mathbf{1}$*) by* a*, we obtain the following possibilities*:

TABLE 2

	a^2	$a \odot b$	b^2
$B_1^{(4)}$	$\mathbf{0}$	$\mathbf{0}$	$\mathbf{0}$
$B_2^{(4)}$	$\mathbf{0}$	$\mathbf{0}$	a
$B_3^{(4)}$	$\mathbf{0}$	$\mathbf{0}$	b
$B_4^{(4)}$	$\mathbf{0}$	a	b
$B_5^{(4)}$	a	a	a
$B_6^{(4)}$	a	a	b

TABLE 3

	a^2	$a \odot b$
$C_1^{(4)}$	$\mathbf{0}$	a
$C_2^{(4)}$	a	a
$C_3^{(4)}$	a	b

c) $\mathbf{0} < a < \mathbf{1} < b$, $b^2 = b$ *and the multiplication* $\odot$ *is determined as in Table* 3;
d) $\mathbf{0} < \mathbf{1} < a < b$, $b^2 = b$, $a \odot b = b$ *and* $D_1^{(4)}: a^2 = a$, $D_2^{(4)}: a^2 = b$;
e) $a \oplus \mathbf{1} = b$ (*i.e.,* $\mathbf{1}$ *and* a *are incomparable*) *and* $E_1^{(4)}: a^2 = a$, $a \odot b = a$, $E_2^{(4)}: a^2 = b$, $a \odot b = b$.

The proofs of these theorems can be obtained by going through all the possibilities, taking into account available restrictions.

Theorems 2.1 and 2.2 show that finite idempotent semirings without additional restrictions form a vast class.

§3. On finite semirings satisfying the cancellation condition

THEOREM 3.1. *The semiring consisting of two elements* $\mathbf{0}$ *and* $\mathbf{1}$ *is the unique finite idempotent semiring satisfying the cancellation condition.*

PROOF. If K is a finite semiring satisfying the cancellation condition, then $\mathbf{1}$ is the maximal element of this semiring (see [1]). And so if $a \in K$, $a \neq \mathbf{1}$, then $a < \mathbf{1}$, whence $a^2 \leqslant a$. Let a be one of the smallest nonzero elements such that $a < \mathbf{1}$. Then this condition implies that $a^2 = \mathbf{0}$ or $a^2 = a$; in any case this contradicts the cancellation condition.

§4. Free idempotent semirings

DEFINITION 4.1. Let K, $\widetilde{K}$ be semirings and $f\colon K \to \widetilde{K}$ be a mapping such that $f(\mathbf{0}) = \mathbf{0}$, $f(\mathbf{1}) = \mathbf{1}$, $f(a \oplus b) = f(a) \oplus f(b)$, and $f(a \odot b) = f(a) \odot f(b)$ for all $a, b \in K$. Then f is called a *homomorphism of the semiring* K into the semiring $\widetilde{K}$.

DEFINITION 4.2. A semiring K_n is called a *free idempotent semiring with n generators* $a_1, a_2, \dots, a_n$ ($a_i \in K_n$), if for any idempotent semiring $\widetilde{K}$

and any elements $b_1, b_2, \dots, b_n \in \widetilde{K}$ there exists a unique homomorphism $f\colon K_n \to \widetilde{K}$ such that $f(a_i) = b_i$, $i = 1, 2, \dots, n$.

THEOREM 4.1. *Up to isomorphism there exists a unique free idempotent semiring with* n *generators, this semiring being isomorphic to the semiring of all finite subsets in* $\mathbb{Z}_+^n$ *(here* $\mathbb{Z}_+ = \{0, 1, 2, \dots\}$*) with the operations of union* $(\oplus)$ *and arithmetic sum* $(\odot)$.

PROOF. Denote by K_n the semiring of all finite subsets in $\mathbb{Z}_+^n$ with the operations indicated above. Suppose $a_i = \{(0, \dots, 0, 1, 0, \dots, 0)\}$ (where 1 occupies the ith place). Then any element of the semiring K_n can be uniquely expressed in the following form:

$$x = x_A = \bigoplus_{(k_1, \dots, k_n) \in A} a_1^{k_1} \odot \dots \odot a_n^{k_n},$$

where A is a finite subset in $\mathbb{Z}_+^n$. Note that $\mathbf{0}$ in K_n corresponds to an empty A and $\mathbf{1}$ coincides with $(0, 0, \dots, 0)$. If $\widetilde{K}$ is another idempotent semiring and $b_1, b_2, \dots, b_n \in \widetilde{K}$, then the unique homomorphism $f\colon K_n \to \widetilde{K}$ which carries a_i into b_i, $i = 1, 2, \dots, n$, is given by the formula

$$f(A) = \bigoplus_{(k_1, \dots, k_n) \in A} b_1^{k_1} \odot \dots \odot b_n^{k_n}.$$

The fact that the free semiring is unique up to isomorphism is evident.

EXAMPLE 4.1. An idempotent semiring in which $(a \oplus b)^2$ does not always coincide with $a^2 \oplus b^2$ (or, which is equivalent, the inequality $a \odot b \preccurlyeq a^2 \oplus b^2$ does not always hold).

The semiring K_2 with $a = a_1$, $b = a_2$ or $a = \{(0; 0), (0; 1)\}$, $b = \{(0; 0), (1; 0)\}$ is such a semiring.

Recall [1] that in an idempotent semiring satisfying the cancellation condition the inequality $a \odot b \preccurlyeq a^2 \oplus b^2$ and the equality $(a \oplus b)^2 = a^2 \oplus b^2$ are always valid. Generally speaking, to check any identity which is valid for any elements of any idempotent semiring, it is sufficient to do so only for the elements $a_1, a_2, \dots, a_n$ of the semiring K_n.

§5. The structure of homomorphisms

DEFINITION 5.1. An equivalence relation $\sim$ in a semiring K is called *compatible with the operations* (of this semiring), if the relations $a_1 \sim a_2$, $b_1 \sim b_2$ ($a_1, b_1 \in K$) imply $a_1 \oplus b_1 \sim a_2 \oplus b_2$ and $a_1 \odot b_1 \sim a_2 \odot b_2$.

If in an idempotent semiring K an equivalence relation $\sim$ compatible with the operations is given, then the quotient set $\widetilde{K} = K/\sim$ (the set of equivalence classes) inherits the structure of an idempotent semiring endowed with the neutral elements $\tilde{\mathbf{0}} = \mathbf{0}/\sim$ and $\tilde{\mathbf{1}} = \mathbf{1}/\sim$ and the operations

$$(a/\sim) \oplus (b/\sim) = (a \oplus b)/\sim, \qquad (a/\sim) \odot (b/\sim) = (a \odot b)/\sim,$$

where $a/\sim$ denotes the equivalence class containing the element a, and the mapping $p\colon K \to \widetilde{K} = K/\sim$ which carries a into $a/\sim$ is an epimorphism of the semirings.

Conversely, let K_1, K_2 be semirings and $f : K_1 \to K_2$ be a homomorphism of these semirings. Then the equivalence relation in K_1 given as follows:

$$a \sim b \iff f(a) = f(b)$$

is compatible with the operations in K_1 and determines a homomorphism $\tilde{f}\colon K_1/\sim \to K_2$ which is an isomorphism of the semiring $K_1/\sim$ on the image of the homomorphism f ($\operatorname{Im} f$). (Obviously, $\operatorname{Im} f$ also has a semiring structure itself.)

In considering homomorphisms of rings, an important role is played by the kernels of homomorphisms. As in the case of rings, one can introduce the kernel of a homomorphism $f\colon K_1 \to K_2$, where K_1, K_2 are semirings:

$$\operatorname{Ker} f = f^{-1}(\mathbf{0}) = \{ a \mid a \in K_1 , f(a) = \mathbf{0} \}.$$

DEFINITION 5.2. An *ideal* in a semiring K is defined as a subset $J \subset K$ such that the following conditions hold:

a) $\mathbf{0} \in J$;
b) $a, b \in J \implies a \oplus b \in J$;
c) $a \in J, b \in K \implies a \odot b \in J$.

Any ideal $J \subset K$ induces (in K) the following equivalence relation compatible with the operations:

$$a \sim b \iff \exists c, d \in J,\ a \oplus c = b \oplus d.$$

We shall denote the semiring $K/\sim$ by K/J and instead of writing $a/\sim$ we shall write $a \pmod J$.

PROPOSITION 5.1. *Let K_1, K_2 be semirings and the mapping $f\colon K_1 \to K_2$ be a homomorphism. Then* $\operatorname{Ker} f$ *is an ideal in the semiring K_1 and the natural homomorphism $\tilde{f}\colon K_1/\operatorname{Ker} f \to K_2$ which carries $a (\operatorname{mod} \operatorname{Ker} f)$ to $f(a)$ is defined.*

It is clear that $\tilde{f}$ also induces an epimorphism $\tilde{f}\colon K_1/\operatorname{Ker} f \to \operatorname{Im} f$. In the case of rings it is an isomorphism. But in the case of semirings this, generally speaking, is not true.

EXAMPLE 5.1. Let K_1, K_2 be idempotent semirings, and let $K = (K_1 \setminus \{\mathbf{0}\}) \times (K_2 \setminus \{\mathbf{0}\}) \cup \mathbf{0}$ with the natural pointwise operations in $(K_1 \setminus \{\mathbf{0}\}) \times (K_2 \setminus \{\mathbf{0}\})$. Now consider the projection $p\colon K \to K_1$ which carries $\mathbf{0}$ to $\mathbf{0}$ and (k_1, k_2) to k_1 for $k_j \in K_j \setminus \{\mathbf{0}\}$, $j = 1, 2$. Then p is a homomorphism of semirings and $\operatorname{Ker} p = \mathbf{0}$, but p is not injective, since $p^{-1}(k_1) = \{(k_1, k_2) \mid k_2 \in K_2 \setminus \{\mathbf{0}\}\}$ for any $k_1 \in K_1 \setminus \{\mathbf{0}\}$.

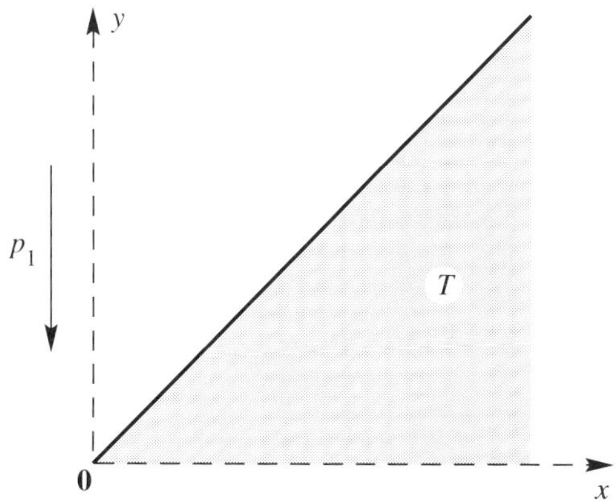

FIGURE 1

Note that if K_1 and K_2 are semirings with division, then K is also a semiring with division. So that the homomorphism $\tilde{f}$ from Proposition 5.1 need not be injective even in the case of semirings with division.

EXAMPLE 5.2. Let $\mathscr{R} = \mathbb{R}^1 \cup \{-\infty\}$ with the operations $\oplus = \max$, $\odot = +$ and the neutral elements $\mathbf{0} = -\infty$, $\mathbf{1} = 0$. Suppose that $\mathscr{R}_+^n = \mathbb{R}^n \cup \{\mathbf{0}\} = (\mathscr{R} \setminus \{\mathbf{0}\})^n \cup \{\mathbf{0}\}$. Introducing the pointwise operations on $\mathscr{R}_+^n$ as in Example 5.1, we see that $\mathscr{R}_+^n$ is a semiring with division.

Now in $\mathscr{R}_+^2$ consider the subsemiring

$$T = \{(x, y) \mid x \geqslant y\} \cup \{0\} \subset \mathscr{R}_+^2.$$

Then T is a semiring satisfying the cancellation condition but without division. The projection $p_1 \colon T \to \mathscr{R}^1$ which carries the pair $(x, y) \in \mathbb{R}^2$ to $x \in \mathbb{R}^1 \subset \mathscr{R}$ and $\mathbf{0}$ to $\mathbf{0}$ is a homomorphism of the semiring T into the semiring $\mathscr{R}$ (see Figure 1).

It is clear that $\operatorname{Ker} p_1 = \{\mathbf{0}\}$; at the same time, $p_1^{-1}(x) = \{(x, y) \mid y \leqslant x\}$ for an arbitrary element $x \in \mathbb{R}^1 = \mathscr{R} \setminus \{\mathbf{0}\}$.

Now we pass to the consideration of semirings with division. Let K_1, K_2 be semirings. If K_1 is a semiring with division and $f \colon K_1 \to K_2$ is a homomorphism, then $\operatorname{Im} f$ is also a semiring with division and $\operatorname{Ker} f = \{\mathbf{0}\}$, but (as we already saw) the mapping f is not necessarily injective. In this case it is convenient to consider, instead of $\operatorname{Ker} f$, the kernel $\operatorname{Ker}^* f$ of the homomorphism

$$f \colon K_1 \setminus \{\mathbf{0}\} \to \operatorname{Im} f \setminus \{\mathbf{0}\},$$

where $K_1 \setminus \{\mathbf{0}\}$ and $\operatorname{Im} f \setminus \{\mathbf{0}\}$ are regarded as groups under multiplication, by putting

$$\operatorname{Ker}^* f = f^{-1}(\mathbf{1}) = \{a \mid a \in K_1,\ f(a) = \mathbf{1}\}.$$

Then the homomorphism which is induced by f on the multiplicative groups (groups with respect to the operation $\odot$)

$$(K_1 \setminus \{\mathbf{0}\}) / \operatorname{Ker}^* f \to \operatorname{Im} f \setminus \{\mathbf{0}\}$$

is an isomorphism by the theorem about homomorphisms of groups. But it remains to clarify what the "multiplicative kernel" $\operatorname{Ker}^* f$ is like in the case when f is a homomorphism of semirings. Now we answer this question.

DEFINITION 5.3. Let K be a semiring with division. A subgroup L of the multiplicative group $K \setminus \{\mathbf{0}\}$ is called *convex*, if $\lambda \odot a \oplus \mu \odot b \in L$ for any $a, b \in L$ and any $\lambda, \mu \in K$ such that $\lambda \oplus \mu = \mathbf{1}$.

The convexity condition can be written in the form $\lambda \odot L \oplus \mu \odot L \subset L$ whenever $\lambda, \mu \in K$ and $\lambda \oplus \mu = \mathbf{1}$. Note that the inclusion $L \subset \lambda \odot L \oplus \mu \odot L$ is obvious in this case. So that in fact the convexity condition can be written as follows: $\lambda \odot L \oplus \mu \odot L = L$ if $\lambda, \mu \in K$ and $\lambda \oplus \mu = \mathbf{1}$.

THEOREM 5.1. *Suppose K_1, K_2 are semirings, K_1 being a semiring with division. Let the homomorphism $f \colon K_1 \to K_2$ be given. Then $\operatorname{Ker}^* f$ is a convex subgroup in $K_1 \setminus \{\mathbf{0}\}$, $((K_1 \setminus \{\mathbf{0}\}) / \operatorname{Ker}^* f) \cup \{\mathbf{0}\}$ has a semiring structure itself and f induces an isomorphism of this semiring (which is actually a semiring with division) on $\operatorname{Im} f$.*

Conversely, if K is an idempotent semiring with division and L is a convex subgroup of the multiplicative group $K \setminus \{\mathbf{0}\}$, then in a natural way the idempotent semiring structure is induced on the set $\widetilde{K} = \{\mathbf{0}\} \cup ((K \setminus \{\mathbf{0}\}) / L)$, so that there is a canonical epimorphism $f \colon K \to \widetilde{K}$ for which $\operatorname{Ker}^* f = L$.

PROOF. a) Suppose the homomorphism $f \colon K_1 \to K_2$, where K_1 and K_2 are semirings, is given, K_1 being a semiring with division, $L = \operatorname{Ker}^* f = f^{-1}(\mathbf{1})$. Now we check the convexity of the subgroup L. If $a, b \in L$, $\lambda, \mu \in K$ and $\lambda \oplus \mu = \mathbf{1}$, then $f(a) = f(b) = \mathbf{1}$ by the definition of L, whence

$$\begin{aligned} f(\lambda \odot a \oplus \mu \odot b) &= f(\lambda) \odot f(a) \oplus f(\mu) \odot f(b) \\ &= f(\lambda) \oplus f(\mu) = f(\lambda \oplus \mu) = \mathbf{1}, \end{aligned}$$

i.e., $\lambda \odot a \oplus \mu \odot b \in L$, as required.

b) Conversely, let L be a convex subgroup of the multiplicative group $K \setminus \{\mathbf{0}\}$, where K is a semiring with division. Now we consider the set $\widetilde{K} = \{\mathbf{0}\} \cup ((K \setminus \{\mathbf{0}\}) / L)$ and show that the operations of the semiring K induce a semiring structure on $\widetilde{K}$.

It is clear that $(a \odot L) \odot (b \odot L) \subset (a \odot b) \odot L$ for all $a, b \in K$. It remains to check that $(a \odot L) \oplus (b \odot L) \subset (a \oplus b) \odot L$ for all $a, b \in K$. If $a = b = \mathbf{0}$ (or, which is the same, $a \oplus b = \mathbf{0}$), then this inclusion is evident, but if $a \oplus b \neq \mathbf{0}$, then it is reduced to the inclusion $\lambda \odot L \oplus \mu \odot L \subset L$, where $\lambda = a \odot (a \oplus b)^{-1}$, $\mu = b \odot (a \oplus b)^{-1}$. Thus the convexity of L guarantees

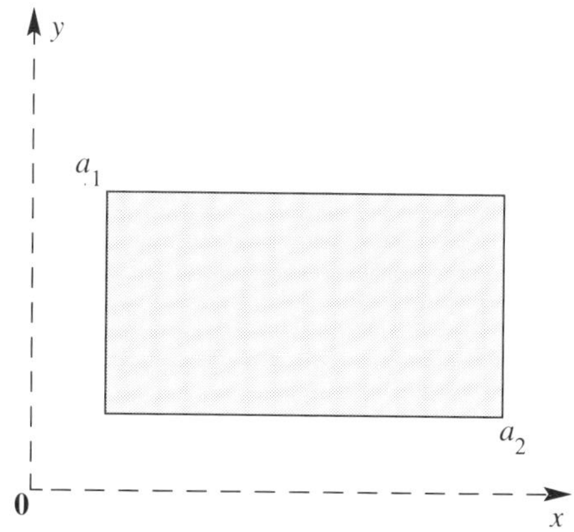

FIGURE 2

that the equivalence relation on $K \setminus \{\mathbf{0}\}$ given by the decomposition into cosets in the subgroup L is compatible with the operation $\oplus$. The theorem is proved.

EXAMPLE 5.3. Suppose $K = \mathscr{R}_+^2 = \mathbb{R}^2 \cup \{\mathbf{0}\}$, L is a subgroup in $\mathbb{R}^2$ with respect to the usual vector addition. Now we clarify the geometric meaning of the convexity of L. Verifying the inclusion $\lambda \odot L \oplus \mu \odot L \subset L$ for $\lambda \oplus \mu = \mathbf{1}$, one can assume that $\lambda, \mu \neq \mathbf{0}$ so that $\lambda, \mu \in \mathbb{R}^2$ (if one of the elements λ, μ coincides with $\mathbf{0}$, then the inclusion is obvious). Let $\lambda = (\lambda_1, \lambda_2)$, $\mu = (\mu_1, \mu_2)$; the equality $\lambda \oplus \mu = \mathbf{1} = (0, 0)$ means that $\max(\lambda_1, \mu_1) = \max(\lambda_2, \mu_2) = 0$, i.e., $\lambda_i \leqslant 0$, $\mu_i \leqslant 0$ and if, for example, $\lambda_1 < 0$, then $\mu_1 = 0$. If $\lambda = \mathbf{1}$, i.e., $\lambda_1 = \lambda_2 = 0$, then the inclusion is obvious as in the case when $\mu = \mathbf{1}$. Hence it is sufficient to assume that $\lambda = (s, 0)$, $\mu = (0, t)$, where $s < 0$, $t < 0$, and the convexity of L means that if $a_j = (x_j, y_j) \in L$, $j = 1, 2$, then we have $(\max\{s+x_1, x_2\}, \max\{y_1, t+y_2\}) \in L$ for all $s < 0$, $t < 0$. Geometrically this means that the rectangle with sides parallel to the coordinate axes and with antipodal vertices a_1 and a_2 is contained in L under the condition $a_1, a_2 \in L$ (Figure 2).

The situation is similar in $\mathscr{R}_+^n$: if L is a subgroup in $\mathbb{R}^n$ with respect to vector addition, then the convexity of L means the following: let $a = (a_1, a_2, \dots, a_n) \in L$ and $b = (b_1, b_2, \dots, b_n) \in L$; then $[c_1, d_1] \times \cdots \times [c_n, d_n] \subset L$, where $c_i = \min\{a_i, b_i\}$, $d_i = \max\{a_i, b_i\}$.

Now it is easy to give a complete description of all convex subgroups in $\mathbb{R}^n$ and the corresponding quotient semirings.

THEOREM 5.2. a) *Convex subgroups in* $\mathbb{R}^n = \mathscr{R}_+^n \setminus \{\mathbf{0}\}$ *are the coordinate subspaces, i.e., subspaces of the form*

$$L = \{x = (x_1, \dots, x_n) \mid x_{i_1} = x_{i_2} = \cdots = x_{i_r} = 0\},$$

where $1 \leqslant i_1 < \cdots < i_r \leqslant n$.

b) *Any quotient semiring of the semiring* $\mathscr{R}_+^n$ *in the multiplicative subgroup* $L \subset \mathscr{R}_+^n \setminus \{\mathbf{0}\}$ *(or, which is the same, any homomorphic image of the semiring* $\mathscr{R}_+^n$*) is isomorphic to* $\mathscr{R}_+^k$ *for some* $k \leqslant n$ *which is uniquely determined.*

PROOF. a) The convexity of a subgroup $L \subset \mathbb{R}^n$ in the sense of Definition 5.3 implies the convexity of L as a subset in $\mathbb{R}^n$ in the usual sense (this follows from Example 5.3). Obviously, the convex subgroup $L \subset \mathbb{R}^n$ is a subspace in $\mathbb{R}^n$. But the convexity in the sense of Definition 5.3 gives additional information about L: if for a collection $\{j_1, \dots, j_l\} \subset \{1, 2, \dots, n\}$ there exist vectors $a^{(1)}, \dots, a^{(l)} \in L$ such that $a_{j_s}^{(s)} \neq 0$, $s = 1, \dots, l$, then L contains the coordinate subspace determined as follows: all coordinates except those with the numbers $j_1, \dots, j_l$ equal zero. Indeed, let us join every $a^{(s)}$ to the point $0 \in \mathbb{R}^n$ by a segment. Then taking the parallelepipeds spanned by these segments and taking the subgroup generated by these parallelepipeds in $\mathbb{R}^n$, we obtain the coordinate subspace corresponding to all nonzero coordinates of the vectors $a^{(s)}$. Thus the first assertion of the theorem is established.

b) The description of quotient semirings of the semiring $\mathscr{R}_+^n$ follows from a) and Theorem 5.1. It remains to prove that the semirings $\mathscr{R}_+^k$ and $\mathscr{R}_+^l$ are nonisomorphic for $k \neq l$. In fact, a stronger assertion is true: $\mathscr{R}_+^k$ and $\mathscr{R}_+^l$ are *nonisomorphic as partially ordered sets.*

First of all note that the canonical partial order on the semiring $\mathscr{R}_+^n$ induces the pointwise partial order on $\mathbb{R}^n = \mathscr{R}_+^n \setminus \{\mathbf{0}\}$: $x = (x_1, \dots, x_n) \leqslant y = (y_1, \dots, y_n)$ if and only if $x_i \leqslant y_i$ for all $i = 1, 2, \dots, n$.

A *ray* in $\mathbb{R}^n$ is a linearly ordered subset $l \subset \mathbb{R}^n$ such that:

(i) $x \in l \implies x \geqslant 0$;
(ii) if $a \in l$ and $0 \leqslant x \leqslant a$, then $x \in l$;
(iii) l is a maximal linearly ordered subset in $\mathbb{R}^n$ satisfying conditions (i) and (ii).

Now we prove that there are exactly n rays in $\mathbb{R}^n$, these rays being the coordinate ones $l_i = \{te_i \mid t \geqslant 0\}$, where $e_i = (0, \dots, 0, 1, 0, \dots, 0)$ (1 occupies the ith place). Indeed, if a vector $a \in l$ has two nonzero coordinates, then not all $x \in [0, a] = \{x \mid 0 \leqslant x \leqslant a\}$ are comparable. Two vectors from l cannot have different nonzero coordinates, since in this case they are noncomparable. Thus $l \subset l_i$ for some $i = 1, 2, \dots, n$. From the fact that l is a maximal subset it now follows that $l = l_i$.

Note that the rays in $\mathbb{R}^n$ are defined only in terms of the partial order and $\mathbb{R}^n$ has exactly n rays. And so the partially ordered sets $\mathbb{R}^k$ and $\mathbb{R}^l$ are nonisomorphic for $k \neq l$. This is sufficient to establish the theorem.

REMARK 5.1. From the second part of the proof it follows that one can define the dimension n of the semiring $\mathscr{R}_+^n$ as the greatest possible number of rays. The same definition is also applicable in the case of any idempotent semiring, when defining rays by conditions (i)–(iii) with the condition $x \geqslant \mathbf{1}$ replacing $x \geqslant 0$ in (i). The dimension obtained is monotone with respect to inclusions.

§6. Completion by ideals

The order relation $\leqslant$ in an idempotent semiring K possesses the following property: for all $a, b \in K$, $\sup\{a, b\}$ exists and is equal to $a \oplus b$. Hence it follows that if $a_1, a_2, \dots, a_n \in K$, then

$$\sup\{a_1, a_2, \dots, a_n\} = a_1 \oplus a_2 \oplus \cdots \oplus a_n.$$

But, whether one can take the supremum of an infinite subset in K (or the "sum" of an infinite series) is unclear. As to the greatest lower bound (infimum) it need not exist even for two-element sets (however, it exists for an idempotent semiring with division, where $\inf\{a, b\} = (a^{-1} \oplus b^{-1})^{-1}$ for $a \neq \mathbf{0}$, $b \neq \mathbf{0}$). But as we shall see in the sequel, one can always imbed any idempotent semiring K into a semiring $\widehat{K}$ which is a *complete lattice* with respect to its partial order (i.e., *every bounded subset* in $\widehat{K}$ *has a supremum and an infimum*). Such a semiring $\widehat{K}$ is given by the general (in lattice theory) construction of *completion by ideals*.

Now we recall this construction (see [2]).

Let K be a partially ordered set in which any two elements $a, b \in K$ have a supremum $\sup\{a, b\} \in K$. An ideal in K is a nonempty subset $J \subset K$ such that the following conditions hold:

a) $a \in J$, $x \in K$, $x \leqslant a \implies x \in J$;
b) $a, b \in J \implies \sup\{a, b\} \in J$.

One can formulate these conditions equivalently as follows: $a \in J, b \in J \iff \sup\{a, b\} \in J$. The *principal ideal* determined by the element $a \in K$ is by definition the ideal

$$K_a = \{x \mid x \in K, x \leqslant a\}.$$

Denote by $\widehat{K}$ the set of all ideals in K. Then $\widehat{K}$ is partially ordered under imbedding and there exists an imbedding $i: K \to \widehat{K}$ assigning to every $a \in K$ the principal ideal K_a. This imbedding is order-preserving.

It is known that if K has a minimal element (in the case when K is an idempotent semiring such an element is $\mathbf{0}$), then $\widehat{K}$ is a complete lattice; to prove this it is only necessary to note that $\inf\{J_\alpha\}_{\alpha \in A} = \bigcap_{\alpha \in A} J_\alpha$ for any set of ideals J_α, and $\sup Q$ is equal to the infimum of all majorants of the set Q.

THEOREM 6.1. *Suppose K is an idempotent semiring and $\widehat{K}$ its completion by ideals. Then $\widehat{K}$ also has an idempotent semiring structure itself and* *the canonical imbedding* $i\colon K \to \widehat{K}$ *is a homomorphism of semirings. This semiring structure in* $\widehat{K}$ *is determined by the following operations*:

$$J_1 \oplus J_2 = \sup\{J_1, J_2\} = \{a \oplus b \mid a \in J_1, b \in J_2\},$$

$$J_1 \odot J_2 = \left\{x \mid x \in K, \exists a_1, \dots, a_n \in J_1, b_1, \dots, b_n \in J_2, x \leqslant \bigoplus_{j=1}^{n} a_j \odot b_j\right\}.$$

Thus, $J_1 \odot J_2$ *is the ideal generated by all elements of the form* $x_1 \odot x_2$, $x_j \in J_j$, $j = 1, 2$, *i.e.,* $J_1 \odot J_2$ *is the smallest ideal containing all such elements.*

PROOF. It is only necessary to check that the multiplication is associative and distributive with respect to addition (the fact that addition is associative is evident). The associativity of multiplication comes from the following assertion: if $J_1, J_2, J_3 \in \widehat{K}$, then $(J_1 \odot J_2) \odot J_3$ and $J_1 \odot (J_2 \odot J_3)$ coincide with the ideal generated by all elements of the form $x_1 \odot x_2 \odot x_3$, where $x_j \in J_j$, $j = 1, 2, 3$. Now we show the multiplication $\odot$ to have the distributivity property with respect to the addition $\oplus$ in $\widehat{K}$. Suppose A, B, C are ideals in K. At first, we prove the following inclusion: $A \odot (B \oplus C) \subset A \odot B \oplus A \odot C$. The inclusion $x \in A \odot (B \oplus C)$ means that $x \leqslant \bigoplus_{j=1}^{n} a_j \odot d_j$, where $d_j \in B \oplus C$, i.e., $d_j \leqslant b_j \oplus c_j$ for some $b_j \in B$, $c_j \in C$.

Since it is possible to multiply inequalities, we obtain

$$x \leqslant \bigoplus_{j=1}^{n} a_j \odot (b_j \oplus c_j) = \left(\bigoplus_{j=1}^{n} a_j \odot b_j\right) \oplus \left(\bigoplus_{j=1}^{n} a_j \odot c_j\right),$$

and it remains to note that the first bracket on the right-hand side of the equality belongs to $A \odot B$ and the second belongs to $A \odot C$.

Now we check the converse inclusion $A \odot B \oplus A \odot C \subset A \odot (B \oplus C)$. If $x \in A \odot B \oplus A \odot C$, then $x \leqslant m \oplus n$, where $m \in A \odot B$, $n \in A \odot C$, i.e.,

$$m \leqslant \bigoplus_{j=1}^{n} a_j \odot b_j, \qquad n \leqslant \bigoplus_{j=1}^{n} a_j \odot c_j$$

for some $a_j \in A$, $b_j \in B$, $c_j \in C$, $j = 1, \dots, n$ (the system $\{a_j\}$ can be considered as the common one for m and n by formally adding terms of the form $a_j \odot \mathbf{0}$ to every sum which estimates m and n). But this implies that

$$m \oplus n \leqslant \bigoplus_{j=1}^{n} a_j \odot (b_j \oplus c_j),$$

so that $m \oplus n \in A \odot (B \oplus C)$ and $x \in A \odot (B \oplus C)$. The desired assertion is established.

Unfortunately completion by ideals does not always lead to a natural result:

it can, for instance, lead to the loss of the cancellation condition and also give a nontrivial extension of a semiring which is already a complete lattice.

EXAMPLE 6.1. Suppose $K = \mathscr{R} = \mathbb{R}^1 \cup \{-\infty\}$. Then ideals in K have one of the following forms:

$$\{-\infty\},\ [-\infty, a),\ [-\infty, a],\ [-\infty, +\infty).$$

Thus besides the maximal ideal $[-\infty, +\infty)$ and the principal ones $a^+ = [-\infty, a]$, more ideals of the form $a^- = [-\infty, a)$ arise so that every point $a \in \mathbb{R}^1$ splits into a pair $a^- < a^+$. When multiplying, one should keep to the following rule of signs:

$$a^- \odot b^- = (a+b)^-, \quad a^- \odot b^+ = (a+b)^-, \quad a^+ \odot b^+ = (a+b)^+.$$

Note, in particular, that $a^- \odot b^- = a^- \odot b^+$ for all $a, b \in \mathbb{R}^1$, so that the cancellation condition, which is satisfied in $\mathscr{R}$, is not satisfied in $\widehat{\mathscr{R}}$.

EXAMPLE 6.2. Suppose $K = \mathbb{Q} \cup \{-\infty\}$, where $\mathbb{Q}$ is the set of all rational numbers. Then $\widehat{K}$ coincides neither with $\mathscr{R}$ nor with $\widehat{\mathscr{R}}$. After completion by ideals, split rational points arise and also single irrational points (besides the maximal ideal K). Thus the cancellation condition does not hold here.

EXAMPLE 6.3. In Example 6.2, by completing we lost the cancellation condition. Let the cancellation condition be satisfied in K, K being a complete lattice. Consider the quotient semiring $\widetilde{K}$. In some cases the semiring $\widetilde{K}$ may be incomplete. Indeed, let $K \subset \mathscr{R}$ have the following form:

$$\begin{aligned} K = \{-\infty\} &\cup \{0 + \mathbb{Z}_+\} \cup \{1 + 2^{-1}\mathbb{Z}_+\} \\ &\cup \{2 + 4^{-1}\mathbb{Z}_+\} \cup \cdots \cup \{n + 2^{-n}\mathbb{Z}_+\} \cup \cdots, \end{aligned}$$

where $\mathbb{Z}_+ = \{0, 1, 2, \dots\}$. The completeness of K is obvious, since $K \cap \mathbb{R}^1$ is discrete. But the semiring of fractions

$$\widetilde{K} = \{-\infty\} \cup \left(\bigcup_{n=1}^{\infty} 2^{-n}\mathbb{Z} \right)$$

consists of $-\infty$ and all dyadic rational numbers and obviously is not a complete lattice.

§7. The kernel theorem in spaces of bounded functions

Suppose K is an idempotent semiring. A set $S \subset K$ is called *bounded*, if it has a majorant, i.e., an element $b \in K$ such that $s \leqslant b$ for any $s \in S$ (here $\leqslant$ is the partial order induced by idempotent addition in K).

DEFINITION 7.1. An idempotent semiring K is called *boundedly complete*, if every bounded subset $S \subset K$ has the least upper bound $\sup S$, i.e., the smallest of all majorants of the subset S.

An idempotent semiring K is called *regular*, if it is *boundedly complete* and possesses the following property: if $\{a_\alpha\}$ is a bounded subset in K and

$\lambda \in K$, then

$$\sup\{\lambda \odot a_\alpha\} = \lambda \odot \sup\{a_\alpha\}. \tag{7.1}$$

In the sequel K will always denote a *regular idempotent semiring*. For example, such is the semiring $\mathscr{R}$.

Assume now that X is an arbitrary set. A function $f\colon X \to K$ is *bounded*, if the set of its values (its range) $\{f(x) \mid x \in X\}$ is bounded. A family of functions $\{f_\alpha\colon X \to K \mid \alpha \in I\}$ is *bounded*, if it is uniformly bounded, i.e., the set $\{f_\alpha(x) \mid x \in X, \alpha \in I\}$ has a majorant in K. In this case the function $\sup_\alpha f_\alpha$ determined by the formula

$$\left(\sup_\alpha f_\alpha\right)(x) = \sup_\alpha\{f_\alpha(x)\}$$

is bounded.

By $\mathscr{B}(X, K)$ we denote the set of all bounded functions on X with values in K. Then $\mathscr{B}(X, K)$ is an idempotent semiring with pointwise operations:

$$(f \oplus g)(x) = f(x) \oplus g(x), \qquad (f \odot g)(x) = f(x) \odot g(x).$$

It is clear that $f \leqslant g$ if and only if $f(x) \leqslant g(x)$ for all $x \in X$. The boundedness of the family $\{f_\alpha \mid \alpha \in I\} \subset \mathscr{B}(X, K)$ in the sense of the semiring $\mathscr{B}(X, K)$ is equivalent to the boundedness introduced above. We have the obvious inclusion $K \subset \mathscr{B}(X, K)$ in which each element $\lambda \in K$ is identified with the function identically equal to λ. The *regularity* of the semiring K implies that of the semiring $\mathscr{B}(X, K)$.

Let X, Y be arbitrary sets. A mapping $A\colon \mathscr{B}(Y, K) \to \mathscr{B}(X, K)$ is said to be a *K-homomorphism*, if the following conditions hold:

(i) $A(0) = \mathbf{0}$;
(ii) $A(f \oplus g) = Af \oplus Ag$, $f, g \in \mathscr{B}(Y, K)$;
(iii) $A(\lambda \odot f) = \lambda \odot Af$, $\lambda \in K$, $f \in \mathscr{B}(Y, K)$.

If $X = Y$, then K-homomorphisms are called *K-endomorphisms*.

EXAMPLE 7.1. Suppose $k \in \mathscr{B}(X \times Y, K)$. Consider the mapping A: $\mathscr{B}(Y, K) \to \mathscr{B}(X, K)$ given by the formula

$$(Af)(x) = \sup_y(k(x, y) \odot f(y)) = \bigoplus\!\!\int k(x, y) \odot f(y)\, dy \tag{7.2}$$

(an *integral K-homomorphism with kernel* $k(x, y)$). Here the last expression is only a notation which emphasizes the analogy with integral operators. It is easy to see that the mapping A is a K-homomorphism (condition (iii) comes from (7.1)).

DEFINITION 7.2. A K-homomorphism $A\colon \mathscr{B}(Y,K) \to \mathscr{B}(X,K)$ is called *normal*, if

$$A\left(\sup_\alpha f_\alpha\right) = \sup_\alpha Af_\alpha \tag{7.3}$$

for any bounded family $\{f_\alpha\} \subset \mathscr{B}(Y, K)$.

It appears that the normality property is exactly the continuity condition that guarantees the validity of kernel theorems of Schwartz's type.

THEOREM 7.1. *Suppose K is a regular idempotent semiring, and the K-homomorphism*

$$A: \mathscr{B}(Y, K) \to \mathscr{B}(X, K), \tag{7.4}$$

where X, Y are arbitrary sets, is given. Then the following conditions are equivalent:

(i) *A is normal*;

(ii) *A is an integral K-homomorphism with kernel $k \in \mathscr{B}(X \times Y, K)$, the kernel k being uniquely determined by the K-homomorphism A.*

PROOF. 1) (ii) $\Rightarrow$ (i). First note that if the bounded function $g: X \times Y \to K$, where X, Y are arbitrary sets, is given, then the analog of Fubini's theorem

$$\sup_{x \in X}\Big(\sup_{y \in Y} g(x, y)\Big) = \sup_{y \in Y}\Big(\sup_{x \in X} g(x, y)\Big) = \sup_{(x, y) \in X \times Y} g(x, y) \tag{7.5}$$

is valid. (This is true in any partially ordered set under the condition that any bounded subset of it has a least upper bound, and follows from the definition of least upper bound.)

Let now $f \in \mathscr{B}(X, K)$, $f = \sup_{\alpha \in I} f_\alpha$, where $f_\alpha \in \mathscr{B}(Y, K)$. If A is an integral K-homomorphism with kernel $k \in \mathscr{B}(X \times Y, K)$, then

$$\begin{aligned}(Af)(x) &= \sup_{y \in Y}(k(x, y) \odot f(y)) = \sup_{y \in Y}\Big(k(x, y) \odot \sup_{\alpha \in I} f_\alpha(y)\Big) \\ &= \sup_{y \in Y}\Big(\sup_{\alpha \in I} k(x, y) \odot f_\alpha(y)\Big) = \sup_{\alpha \in I}\Big(\sup_{y \in Y}(k(x, y) \odot f_\alpha(y))\Big) \\ &= \Big[\sup_{\alpha}(Af_\alpha)\Big](x)\end{aligned}$$

(the regularity identity and relation (7.5) were used to obtain this series of equalities). Hence the normality of A is proved. Using (7.1) and (7.5), one also can verify the algebraic requirements from the definition of the K-homomorphism.

2) Now we prove that the kernel k can be uniquely recovered from the integral K-homomorphism A. For this it is sufficient to note that

$$k(x, y) = (A\delta_y)(x), \tag{7.6}$$

where $y \in Y$ and $\delta_y \in \mathscr{B}(Y, K)$ is defined as a "δ-function" with support at the point y, i.e.,

$$\delta_y(z) = \begin{cases} \mathbf{1} & \text{for } z = y, \\ \mathbf{0} & \text{for } z \neq y. \end{cases}$$

3) Finally, we prove that (i) implies (ii). Let the normal K-homomorphism A (see (7.4)) be given. Now we construct the function $k: X \times Y \to K$ by formula (7.6) and prove that this function k is bounded. Note, first of all, that the additivity property of the K-homomorphism A implies its monotonicity, i.e., $f \leqslant g \Rightarrow Af \leqslant Ag$. Since $\delta_y \leqslant \mathbf{1}$ for any $y \in Y$, we have $k(x, y) \leqslant (A\mathbf{1})(x)$ and it remains to use the fact that $A\mathbf{1} \in \mathscr{B}(X, K)$.

Finally, we check that A is in fact given by formula (7.2). Note that for any function $f \in \mathscr{B}(Y, K)$ the relation

$$f = \sup_{y \in Y}(f(y) \odot \delta_y)$$

holds, whence by the normality of A

$$\begin{aligned}(Af)(x) &= \sup_{y \in Y}(A(f(y) \odot \delta_y))(x) = \sup_{y \in Y}((A\delta_y)(x) \odot f(y)) \\ &= \sup_{y \in Y}(k(x, y) \odot f(y)).\end{aligned}$$

The proof is complete.

REMARK 7.1. The regularity condition of the semiring K is natural in the context of Theorem 7.1, since bounded completeness is necessary in order to define integral K-homomorphisms with arbitrary bounded kernel, and the additional regularity condition (see (7.1)) simply means that the operator of multiplication by $\lambda \in K$ in "one-dimensional" space $K \sim \mathscr{B}(\mathrm{pt}, K)$, where pt is a one-point set, is normal.

REFERENCES

1. P. I. Dudnikov and S. N. Samborskii, *Endomorphisms of semimodules over semirings with an idempotent operation*, preprint 87.48 Math. Inst. Acad. Sci. Ukrain. SSR, 1987; Revised version: Izv. Akad. Nauk SSSR Ser. Mat. **55** (1991), 93–109; English transl. in Math. USSR-Izv. **38** (1992).

2. G. Birkhoff, *Lattice theory*, Amer. Math. Soc., Providence, R. I., 1977.

3. V. N. Kolokoltsov and V. P. Maslov, *The general form of endomorphisms in the space of continuous functions with values in a numerical semiring with idempotent addition*, Dokl. Akad. Nauk SSSR **295** (1987), no. 2, 283–287; English transl., Soviet Math. Dokl. **36** (1988), no. 1, 55–59.

4. ______, *Idempotent calculus as the apparatus of optimization theory*, I, Funktsional. Anal. i Prilozhen. **23** (1989), no. 1, 1–23; II, **23**, no. 4, 53–62; English transl. in Functional Anal. Appl. **23** (1989).

DEPARTMENT OF MATHEMATICS, NORTHEASTERN UNIVERSITY, BOSTON, MA 02115, USA, AND CENTER FOR OPTIMIZATION AND MATHEMATICAL MODELING, INSTITUTE OF NEW TECHNOLOGIES, KIROVOGRADSKAYA UL., 11, 113587, MOSCOW, RUSSIA

Translated by I. PETRENKO

ADVANCES IN SOVIET MATHEMATICS
Volume 13, 1992

Nonlinear Semigroups and Infinite Horizon Optimization

S. YU. YAKOVENKO AND L. A. KONTORER

§1. Introduction

1.1. Economics. In this paper we make an attempt to discuss an approach to infinite horizon stationary optimization problems, based on the notion of (nonlinear) Bellman operators.

These problems are not merely of academic interest, since they constantly appear in the normative theory of growth and capital accumulation in mathematical economics. Speaking rather vaguely, the normative theories describe economic dynamics as follows. A family of *feasible paths*, or *feasible trajectories*, describing a possible dynamics of a system is defined (either explicitly or implicitly). Usually this is done by using so-called *technological restrictions* (e.g., balance equations and capacity limitations are to be satisfied independently of the adopted economic policy). Next, a description of a *criterion*, or of the economic goals pursued by the system under consideration is given. Finally it is assumed (also in more or less explicit form) that the system behaves *rationally*, or optimally. This means that the *true* trajectory provides an optimum to the criterion over all feasible paths starting at the same initial point.

The criterion is a scalar-valued functional defined on all feasible paths (we do not consider here the multicriterial case). Usually it takes the form of a sum or an integral of a certain expression (referred to as the *utility*) over the trajectory. But such a form immediately brings under consideration an important parameter which has been in shadow up to now. This parameter is the *time horizon* (otherwise referred to as the planning horizon).

If this parameter is assumed to take a certain *finite* value, then no additional mathematical difficulties arise, since the finite sums/integrals are quite well defined, thus yielding the correct criterion of optimization. But economic difficulties appear. Their nature can be explained using the consumption/investment dichotomy.

1991 *Mathematics Subject Classification.* Primary 20M20, 49L20; Secondary 90A16.

The latter is the model in which the profit obtained at each step must be divided into the immediate consumption part and the one to be invested to increase the productive capacities of the system, resulting in the increase of future profits. The criterion is integral consumption.

Given a certain *finite* planning horizon, the optimal trajectories of the model (under reasonable assumptions) behave as follows. At earlier stages, there is usually an optimal proportion between consumption and investment which is more or less precisely maintained along the optimal paths. But this proportionality no longer holds when the model is close to the beginning or the end of the time interval. The starting segment is greatly influenced by the initial state, and since this state is prescribed, there is no problem in such a deviation, at least from the point of view of methodology. Completely different is the other extremity, towards the end of the planning horizon. The system's behavior during the final steps follows the "*après nous le déluge*" pattern: the investment almost vanishes.

Clearly, such a property of the model is quite reasonable within the framework of the finite horizon optimization, since the decrease in investments would result in shrinking future profits, but this future is beyond the planning horizon. Apparently, such features of the model have to be avoided somehow. There is a series of tricks for this. The simplest one is to disregard the model as soon as its behavior begins to be influenced by this terminal effect, but this is not so easy in the multidimensional case with more than two alternatives. Some other approaches were developed in different particular cases, but they rely heavily on certain properties of the model (convexity, etc.), and require a thorough preliminary investigation of properties of all the finite extremals; the most illustrative example is turnpike theory.

Instead of finite horizon problems, one may consider infinite horizon extremals. But, in contrast to the finite case, the infinite sums and improper integrals sometimes (and even very often) happen to diverge. Using a kind of regularization procedure, one can try to transform the criteria to something convergent, but the justification of such manipulations hardly goes beyond the heuristic level.

An alternative strategy consists in introducing certain partial orders on all the possible cost flow paths corresponding to all feasible paths (the cost flow path is the sequence of scalars $c_T \in \mathbb{R}$, $T \in \mathbb{N}$ or $T \in \mathbb{R}$, where c_T is the value taken by the criterion at the initial T-step segment of the trajectory). Unfortunately, such partial orders (or, more correctly, binary relations) are not very naturally introduced[1], and the most popular among them are nontransitive (a short survey of this theory is given below).

As an immediate consequence of this nontransitivity the *existence problem* arises: does there exist "best" elements with respect to those orders?

[1]For example, try to compare the following three cost flow paths:

$$0, 2, 0, 2, 0, 2, \dots, \qquad 2, 0, 2, 0, 2, 0, \dots, \qquad 1, 1, 1, 1, 1, 1, \dots.$$

(Here "best" means either majorizing or nonmajorizable.) It often happens that some additional assumptions are necessary to prove the existence theorems. The convexity (or strict convexity) assumption is most important among them, and sometimes the analysis goes along the lines of turnpike theory.

Therefore the economic problem as it was just stated is *to develop a dynamic optimization theory independent of the choice of the planning horizon.*

1.2. Mathematics. Stationary dynamic optimization problems in mathematics usually appear in one of the following frameworks.

The discrete time problem. Let X be the phase space of the model. It can be either a subset of a Euclidean n-space, or a discrete set, or even an abstract topological space. The points of this set represent different states of the system. We assume that the model is deterministic, so the information about the states is complete and precise.

A trajectory or path is a sequence of states $\mathbf{x} = \{x_t \in X : t = 0, 1, 2, \dots\}$. Therefore the *time variable* t is assumed to run over only nonnegative integer values.

Feasible paths are usually defined recursively: there exists a multivalued map $F : X \to 2^X$ and $\mathbf{x}$ is feasible if and only if $x_{t+1} \in F(x_t)$ for all t (the system is stationary, so F is independent of t).

The criterion of optimization in the most general form is the sum

$$\mathscr{B}(\mathbf{x}) = \sum_t b(x_t, x_{t+1}),$$

where $b(\cdot, \cdot) : X \times X \to \mathbb{R}$ is the utility function: the value $b(x, y)$ is associated with the utility of transition from the state x to the state y in one step (this function is also time independent). From now on we will call $b(\cdot, \cdot)$ the *transition function.*

These data allow us to pose the stationary dynamic optimization problem in discrete time,

$$\mathscr{B}(\mathbf{x}) \to \max, \qquad x_0 = a \in X,$$

where a is a prescribed initial state, and the maximum is taken over all feasible paths starting at a.

In order to simplify the construction, we shall incorporate the feasibility restrictions into the transition function by setting $b(x, y) = -\infty$ if $y \notin F(x)$ (the infinite penalty). Clearly, we shall pay for such a simplification by imposing certain conditions of semicontinuity on b, but all the constructions will become more transparent. Without further mention, we shall make the following

SEMICONTINUITY ASSUMPTION. *The phase state is a separable topological space, and the transition function $b(\cdot, \cdot)$ is upper semicontinuous, that is, its hypograph $\{(x, y, u) \in X \times X \times \mathbb{R} : u \leqslant b(x, y)\}$ is a closed subset.*

In fact, the functional $\mathscr{B}$ is still undefined, since we did not point out the upper limit for the summation. If $N \in \mathbb{N}$ is a finite integer, then we can introduce the functional

$$\mathscr{B}_N(\mathbf{x}) = \sum_{t=0}^{N-1} b(x_t, x_{t+1}).$$

Replacing $\mathscr{B}$ by $\mathscr{B}_N$, we obtain a well-posed finite horizon optimization problem of a common type.

Proposition. *Under the semicontinuity assumption if the phase state is compact and the horizon N finite, then the optimization problem*

$$\mathscr{B}_N(\mathbf{x}) \to \max, \qquad x_0 = a$$

always possesses a solution.

The difficulties arise when defining something like $\mathscr{B}_\infty$.

The above finite horizon problem admits a natural generalization. Let $f: X \to \mathbb{R}$ be any upper semicontinuous function. Then we may add to the criterion the *terminant* $f(x_N)$. The problem thus obtained will be of the form (we write the unabridged expression)

$$\sum_{t=0}^{N-1} b(x_t, x_{t+1}) + f(x_N) \to \max, \tag{1.1}$$

$$x_t \in X, \quad t = 0, 1, \ldots, N, \quad x_0 = a \in X. \tag{1.2}$$

A solution to the problem is an $(N+1)$-tuple from X^{N+1}. The set of all solutions will be denoted by $\operatorname{extr}_N(b, f)$. The above proposition holds also for this case.

This generalization points out another specific aspect of the infinite horizon case. Indeed, the latter does not admit any terminants besides almost meaningless ones like $f(\lim_{t\to\infty} x_t)$. At the same time if we think of the infinite horizon optimization problem as the limit (in some sense) of regular finite horizon problems, the terminants may occur in the latter ones. Thus the "right" definition has to keep some traces of a terminant.

The continuous time case: variational problems. We have given the detailed description of the framework of the discrete time case, since all the phenomena analyzed in the current paper can be studied within it. Nevertheless the continuous time case is also very important.

Now we assume that the phase space X is endowed with the structure of a linear space or at least is a differentiable manifold. In this case we may define and consider absolutely continuous trajectories of the form $x: \mathbb{R} \supseteq \Delta \ni t \mapsto x(t) \in X$.

The role of the transition function is played by a Lagrangian function $L(x, v)$, which defines the integral functional

$$\int_0^T L(x(t), \dot{x}(t))\, dt, \qquad t < \infty,$$

to which an optional term $f(x(T))$ may be added. A variational problem consists in finding the maxima of this criterion over all admissible paths starting at a certain point $a \in X$.

The continuous time version leads to greater technical difficulties with the existence problem for the finite horizon case. Usually one requires the concavity of L in v together with relatively rapid decrease to $-\infty$ as $\|v\| \to \infty$. We shall treat this case in its turn.

The continuous time case: *optimal control problems.* The variational problem as it was stated above admits no constraints. Those are generally introduced within the framework of control theory. Let the phase space X be again endowed with a smooth structure and suppose that there is a controlled differential equation on X of the form

$$\dot{x} = F(x, u), \qquad x \in X, \ u \in U,$$

where U is a certain topological space of admissible controls.

This control problem defines admissible solutions, which satisfy the above equation almost everywhere. The criterion for such problems usually takes the form

$$\int_0^T g(x(t), u(t))\, dt + f(x(T)).$$

All the remarks made beforehand refer as well to this case. Actually, the Lagrange function introduced formally as

$$L(x, v) = \sup_{u \in U}\{g(x, u) : F(x, u) = v\}, \qquad \sup \emptyset = -\infty,$$

permits to reduce the optimal control problem to a variational one, if one allows functions taking the value $-\infty$.

We do not dwell on the technical matters relevant to the continuous time case for two reasons. First, these difficulties are not the point of our analysis, so we may always assume any additional regularity of the problem, referring the reader to any source. The second reason is that below we suggest a generalization of the notion of continuous time stationary dynamic optimization problem which includes both possibilities. Of course, some new questions immediately arise, but they will be treated elsewhere.

1.3. Structure of the paper. The paper is organized as follows. First we analyze some heuristic approaches which lead to *a priori* properties of infinite extremals, if they exist. It turns out that different arguments lead to the same functional equation; in the discrete time case (the one we are dealing with almost everywhere throughout the paper), it is of the form

$$\sup_{x \in X}\{b(x, y) + f(y)\} = \lambda + f(x). \tag{BE}$$

Here b is the transition function, $f: X \to \mathbb{R}$ is an unknown function and λ is a certain unknown scalar.

Of course, there is nothing new in this equation, which is simply a discrete version of the stationary Bellman equation. Its solvability was studied in different situations in different sources.

But the left-hand side of this equation may be considered as the result of application of a nonlinear operator, which we suggest to call the *Bellman operator* with kernel b, to the function f. Hence (BE) becomes something like the equation for the eigenfunctions of an integral operator. And this is not merely an analogy. In fact, there is a natural algebraic structure associated with (BE), which is similar to the standard linear one except that there is no subtraction operation inverse to addition (lacking the appropriate term and wishing to stress the analogy, we call this structure "linear" in quotation marks, extending this agreement to all other terms relevant to the structure). The Bellman operators preserve the "linear" structure, and (BE) indeed becomes the spectral equation with respect to it.

This structure was also known before (see the survey section of the paper). "Spectral" properties are extensively studied in other papers constituting this volume. The key point was to link both theories with infinite horizon optimization. As a result, we have developed a kind of nonlinear (though "linear") representation of a dynamic optimization problem in a function space.

The "eigenfunctions" of the Bellman operator generate infinite paths which are natural candidates for being called infinite extremals. We analyze their properties and study all the solutions of (BE). It turns out that they are closely related to limit sets of a dynamical system in the space of the abovementioned "linear" representation.

Some points of the construction can be clarified if we omit the initial state a from the formulation (1.1), (1.2), replacing it by the periodicity condition

$$x_0 = x_N.$$

An axiomatic theory is worthless if it gives no information about already existing problems. One section of the paper relates the axiomatic definition of infinite extremals via "eigenfunctions" to the classical definitions of catching up, optimal, and overtaking trajectories. The main point is that if there are infinite extremals in the classical sense, then they are necessarily generated by some "eigenfunctions", but not vice versa. Perhaps somewhat presumptuously we suggest arguments which show an artificial nature of the classical definitions, and speak in favor of more invariant Bellman-based constructions.

Another point is to highlight the nature of the convexity assumptions specific of mathematical economics from the point of view of the Bellman calculus. We show that the dynamical system in the representation space associated with concave transition functions is dissipative: there is a unique point which is the limit set for all the orbits of the representation. The spectral properties for the concave case are also much simpler.

As for the continuous time case, we associate with any sufficiently regular optimization problem a continuous time dynamic system in the representation case. In fact, this system is a generalization of the evolution semigroup for the Hamilton-Jacobi equation corresponding to the variational problem. We prove spectral properties of the semigroup and discuss some open questions.

The roots and sources of the theory presented here are dispersed among many papers and books, some of them about 30 years old. We made an attempt to translate some known facts and theorems into the language of "linear" semigroups. Since the translation cannot precede the vocabulary, we violated the existing scientific tradition by placing the survey section at the end of the paper. Needless to say that our list had to be incomplete both in the mathematical and the economical parts.

This paper existed in draft form since the time the short communication was submitted for publication (1988). Recently one of the authors (S. Y.) had a chance to meet professor Arie Leizarowitz (The Technion, Haifa) and to get acquainted with series of his very illuminating papers on this subject. Many assertions proved in the current article were proved by Leizarowitz using sometimes similar, sometimes different techniques. Nevertheless we decided to preserve the initial structure of the paper in order to show how the entire area may be developed from a different point of view.

1.4. Acknowledgments. During the preparation of this paper and while developing the apparatus, we enjoyed many beneficial conversations with different people. We are grateful to S. N. Samborskiĭ and V. N. Kolokoltsov for discussions concerning idempotent analysis, A. M. Rubinov, V. Z. Belen′kiĭ, V. M. Polterovich and V. D. Matveenko for numerous opportunities to talk about the economical questions underlying the subject. Talks with Zvi Artstein and Arie Leizarowitz helped us to look at the problems from a different point of view. V. P. Maslov and A. D. Tsvirkun encouraged us and made this research possible. We are very grateful to all of them.

§2. Motivations. Three problems with one solution

2.1. Preliminaries. As we already mentioned in the introduction, there are several possible approaches to definition of the notion of infinite extremals. Usually this is done by using certain binary relations on the space of cost flows, which is the space $\mathbb{R}^{\mathbb{N}}$ of infinite sequences of the form $\mathbf{c} = \{c_0, c_1, \dots, c_t, \dots\}$, $c_t \in \mathbb{R}$. The important example is the binary relation in the space of cost flows, defined as

$$\{c_t\} = \mathbf{c} \geqslant \mathbf{c}' = \{c_t'\} \iff \forall \varepsilon > 0, \exists N = N(\varepsilon) \in \mathbb{N} : \forall t > N, \ c_t > c_t' - \varepsilon.$$

A stronger binary relation is the Pareto partial order on $\mathbb{R}^{\mathbb{N}}$, and an intermediate concept is the asymptotical Pareto order

$$\mathbf{c} \geqslant \mathbf{c}' \iff \exists N < \infty : \forall t > N, \ c_t \geqslant c_t'.$$

Each partial order in the space of cost flows generates the corresponding partial order on infinite trajectories $X^{\mathbb{N}}$: there is defined the *cost flow map*

$$\mathfrak{C}: \mathbf{x} \mapsto \mathbf{c} = \mathfrak{C}(\mathbf{x}) : \quad c_N = \sum_{t=0}^{N-1} b(x_t, x_{t+1}),$$

and we may put

$$\mathbf{x} \geqslant \mathbf{x}' \iff x_0 = x_0' \text{ and } \mathfrak{C}(\mathbf{x}) \geqslant \mathfrak{C}(\mathbf{x}')$$

for any relation $\geqslant$ on $\mathbb{R}^{\mathbb{N}}$.

Other possible binary relations on the space of cost flows are given below. Here we only point out that L. Stern's paper [1] gives a compendium of almost all logically possible definitions together with some relationships among them.

After a certain binary relation $\geqslant$ on the space of infinite trajectories is introduced, it becomes possible to speak of 'maximal' elements. By this one usually means those which majorize (in the sense of the relation $\geqslant$) all the other trajectories starting at the same point. Another possibility is to consider nonmajorizable trajectories, i.e., those which cannot be majorized by any path with the same initial point. This alternative doubles the number of possible definitions.

Further development of these ideas leads to the important notion of overtaking criterion suggested by D. Gale [30] and C. C. von Weizsacker [31]. This approach will be discussed somewhat later in its relationship to the concept of optimality suggested in §§3, 4.

Such a degree of arbitrariness in definitions leads to some doubts. Moreover, soon one discovers the fact that, due to nontransitivity of many relations defined in this way (this refers especially to nonmajorizable elements) and lack of antisymmetry (both possibilities $\mathbf{x} > \mathbf{y}$ and $\mathbf{y} > \mathbf{x}$ may occur simultaneously) these "extremals" do not always exist.

Thus we are led to an attempt to look for other types of definitions, based on different ideas. We shall describe three different approaches leading to the same functional equation (BE).

2.2. Infinite series maximization. Suppose that for some unknown reasons the series

$$S(\mathbf{x}) = \sum_{t=0}^{\infty} b(x_t, x_{t+1})$$

is so good that it generates a reasonable criterion for optimization, that is, it either diverges to $-\infty$ for "bad" trajectories $\mathbf{x}$, or converges and, moreover, the optimal value

$$f(x) = \max_{\mathbf{x}\,:\,x_0=x} S(\mathbf{x}) \tag{2.1}$$

is a well-defined semicontinuous function. Since we are dealing with heuristics, we do not discuss conditions guaranteeing such a behavior.

In such a situation it is clear that the function $f(\cdot)\colon X \to X$ must satisfy the condition

$$\forall x \in X, \quad \max_{y\in X}\{b(x, y) + f(y)\} = f(x). \tag{2.2}$$

Indeed, if the trajectory $\mathbf{x}$ with $x_0 = x$, $x_1 = y$ yields the maximum in the definition of $f(x)$, then the trajectory $\mathbf{y}$ with $y_t = x_{t-1}$ does so with respect to $f(y)$, and vice versa.

Unfortunately, the case of convergence is quite unstable: if by chance the transition function $b(\cdot\,,\cdot)$ generates such a converging criterion, then for any $\lambda \neq 0$, $\lambda \in \mathbb{R}$ the function $b' = b + \lambda$ yields the series S' which is:

— diverging to $-\infty$ for any $\mathbf{x}$ if $\lambda < 0$, or
— possibly converging for certain paths, but diverging to $+\infty$ for those which were optimal with respect to the initial criterion, if $\lambda > 0$.

But this instability also suggests a remedy: given an arbitrary transition function b, one may expect that its average growth rate along "optimal paths" is equal to a certain constant value $\lambda \in \mathbb{R}$, which could be subtracted in order to obtain the above convergence. Substituting this modified transition function $b - \lambda$ into (2.2), we get a somewhat more universal equation

$$\forall x \in X, \quad \max_{y\in X}\{b(x, y) + f(y)\} = \lambda + f(x). \tag{2.3}$$

Now we may forget all the assumptions made about convergence and look at (2.3) as the equation for the unknown function f and the unknown value λ, having in mind that if all the assumptions hold, then the result must satisfy (2.3).

2.3. Dynamic programming. Another attempt is based on the possibility of adding a terminant to the criterion of optimization. Suppose that instead of the finite horizon problem

$$\sum_{t=0}^{N-1} b(x_t, x_{t+1}) \to \max, \qquad x_0 = \text{fixed},$$

we are solving the problem

$$\sum_{t=0}^{N-1} b(x_t, x_{t+1}) + f(x_N) \to \max, \qquad x_0 = \text{fixed}, \tag{2.4}$$

with a certain function f. The idea is to find a terminant f such that the solution of the problem (2.4) would be independent of N in the following sense: if $N' > N$ is a new horizon and $\{x_t\}_{t=0}^{N'}$ is the corresponding solution, then the initial segment $\{x_t\}_{t=0}^{N}$ is the solution to the problem with horizon N.

A regular method to solve problems like (2.4) is provided by the dynamic programming principle. One must construct the sequence of functions defined recursively:

$$f_0(x) = f(x), \qquad f_{s+1}(x) = \sup_{y \in X}\{b(x, y) + f_s(y)\}, \quad s = 0, 1, 2, \ldots, N-1, \tag{2.5}$$

and afterwards construct the sequence of solutions

$$x_{t+1} \in \operatorname*{Arg\,max}_{y \in X}\{b(x_t, y) + f_{N-t-1}(y)\}, \qquad t = 0, 1, \ldots, N-1. \tag{2.6}$$

The desired independence of the solution of N means that the process (2.6) is in fact independent of N. It would be so if all the functions f_s were independent of the index s or at least differed only by some constants. The latter condition holds if there is a certain $\lambda \in \mathbb{R}$ such that $f_0 = f$ satisfies (2.3) together with this λ. Thus to find an appropriate terminant means to solve (2.3).

2.4. Symmetries. The last approach is based upon the idea of transformations of an optimization problem. Let $b: X \times X \to \mathbb{R}$ be a transition function. Along with b we may consider another transition function of the form

$$b'(x, y) = b(x, y) + g(y) - g(x) + \mu, \tag{2.7}$$

where $g: X \to \mathbb{R}$ is a continuous function, and μ is an arbitrary scalar.

The optimization problem with transition function b' is equivalent to the initial one in the following sense:

(1) the two problems

$$\sum_{t=0}^{N-1} b(x_t, x_{t+1}) + f(x_N) \to \max, \qquad x_0 = \text{fixed},$$

$$\sum_{t=0}^{N-1} b'(x_t, x_{t+1}) + (f - g)(x_N) \to \max, \qquad x_0 = \text{fixed},$$

possess the same solutions for all initial values;

(2) if we consider either the fixed endpoints or periodic problems with the boundary conditions correspondingly $x_0 = \text{fixed}$, $x_N = \text{fixed}$ or $x_0 = x_N$, then the replacement of b by b' does not change the extremals at all.

This is so because b' differs from b by the *full difference* (the discrete time analog of the full derivative), and the addition of the constant to the transition function never changes the extremals.

The main idea related to the notion of equivalence is the idea of invariance: equivalent problems must have the same properties. In our case this is to be understood as the claim that *the equivalent transition functions generate the same infinite extremals.* Thus it may prove useful to look for another transition function which would be equivalent to the initial one and at the same time would be simpler to analyze.

Such functions do exist indeed.

DEFINITION 2.1. The transition function $b(x, y)$ is called *good nonpositive*, if

$$\forall x, y \in X, \quad b(x, y) \leqslant 0, \quad \text{and} \quad \forall x \in X, \exists y \in X: \quad b(x, y) = 0.$$

Another way to say that b is good nonpositive is to write

$$\forall x \in X, \quad \max_{y \in X} b(x, y) = 0.$$

Infinite extremals for a good nonpositive transition function are easy to define.

PROPOSITION 2.1. *If $b(x, y)$ is good nonpositive, then*:

1. *For any infinite trajectory* $\mathbf{x}$ *the series*

$$\sum_{t=0}^{\infty} b(x_t, x_{t+1}) \tag{2.8}$$

monotonically converges either to a nonpositive value or to $-\infty$.

2. *For any initial point $x_0 \in X$ there exists at least one infinite trajectory* $\mathbf{x}$ *starting at x_0 such that the series* (2.8) *is identically zero.*

COROLLARY 2.1. *The optimization problem*

$$\sum_{t=0}^{\infty} b(x_t, x_{t+1}) \to \max$$

always admits a solution provided that b is good nonpositive.

A trajectory $\mathbf{x}$ *solves this problem if and only if*

$$b(x_t, x_{t+1}) = 0 \left(= \max_{y \in X} b(x_t, y) \right).$$

Thus the invariance principle leads us to the following task: *to find a good nonpositive function equivalent to a given one.*

The latter problem means that we must solve the equation

$$\forall x, \quad \max_{y \in X} \{b(x, y) + g(y) - g(x) + \mu\} = 0,$$

which is equivalent to (2.3) with $\lambda = -\mu$.

Therefore we see that to find a transformation of the form (2.7) taking the initial transition function b to a certain good nonpositive b', again one must solve the same functional equation.

A natural question that arises here is whether other transformations of the optimization problem, besides those of type (2.7), are possible. At least one such transformation also exists. Namely, let $H: X \to X$ be a one-to-one continuous mapping (the state variable change). Then this change of variable yelds a new transition function

$$\widetilde{b}(x, y) = b(H(x), H(y)). \tag{2.9}$$

Clearly, such a transformation adds nothing new to our understanding of the initial problem: the trajectory $\mathbf{x}$ solves the new problem if and only if $H(\mathbf{x})$ solves the initial one, and vice versa (if X is a discrete compact set, i.e., a graph, then such a transformation corresponds to renumbering its vertices).

But these two types of transformation are in a sense the only possibilities: there is a natural symmetry group associated with optimization problems of this kind, and this group is generated by transformations of the form (2.7) and (2.9). A more precise formulation is given below.

2.5. Summary. The above considerations taken together lead to the following conclusions.

1. *Each transition function* $b(x, y)$ *satisfying the semicontinuity assumption, gives rise to the Bellman equation* (2.3).

2. *Any semicontinuous solution* $(\lambda, f(\cdot))$ *of the Bellman equation generates infinite extremals by means of the recurrent formula*

$$x_{t+1} \in \operatorname*{Arg\,max}_{y \in X}\{b(x_t, y) + f(y)\}. \tag{2.10}$$

Using this formula, one can construct an infinite extremal starting at any point $x_0 \in X$.

3. *If no other data except the transition function are provided, then there is no means to distinguish between different solutions of the Bellman equation.*

4. *When studying optimization problems of this kind, one has to pay attention to the invariance of all procedures and definitions with respect to transformations of the form* (2.7), (2.8).

REMARK. Note that condition (2.10) includes only the "first differences", unlike the discrete time counterpart of the Euler-Lagrange equations

$$x_t \in \operatorname*{Arg\,max}_{y \in X}\{b(x_{t-1}, y) + b(y, x_{t+1})\}, \qquad t = 1, \dots, N-1, \tag{2.11}$$

which is the necessary condition for optimality involving "second order" differences. In other words, the behavior of an infinite extremal depends only on its initial state (though this state might not determine the path uniquely).

§3. Glossary for the "linear" algebra

3.1. Algebra. In this section we introduce the structure of idempotent semimodule on the space of extended real-valued functions on X and list all necessary definitions. Since this subject is extensively covered by the other papers of the present volume, our exposition will be as short as possible.

The main semiring of extended reals is denoted by $\mathscr{R} = \mathbb{R} \cup \{-\infty\}$. This semiring is endowed with the two binary operations $\oplus$ and $\odot$:

$$\forall a, b \in \mathscr{R}, \qquad a \oplus b = \max(a, b), \quad a \odot b = a + b.$$

The metric on $\mathscr{R}$ is defined as $\operatorname{dist}(a, b) = |\exp a - \exp b|$ ($\exp(-\infty) = 0$). This metric generates the corresponding topology on $\mathscr{R}$ endowing it with a topological ordered semiring structure. Note that both operations are commutative and associative, with "addition" $\oplus$ being distributive with respect to "multiplication" $\odot$. Unfortunately, there is no "subtraction" in $\mathscr{R}$.

These structures are inherited by the space $\mathscr{R}(X)$ of $\mathscr{R}$-valued functions on X; the latter bearing thus acquires the structure of a semimodule over the semiring $\mathscr{R}$:

$$\forall f, g \in \mathscr{R}(X), \ \lambda \in \mathscr{R} \qquad (f \oplus g)(x) = f(x) \oplus g(x), \quad (\lambda \odot f)(x) = \lambda \odot f(x).$$

There is a subspace in $\mathscr{R}(X)$ consisting of continuous functions that never assume the value $-\infty$. We denote it in the usual manner by $C(X)$.

The semimodule and semiring structures were independently introduced in many works on combinatorial mathematics and graph theory; see, for example, the paper by F. Gondran [8] and references there, [28], etc. Recently this and similar structures were deeply investigated by V. P. Maslov, S. N. Samborskiĭ, V. N. Kolokoltsov e.a. [9, 10]. To make the analogy with the regular ring/module structure more transparent, we shall use terms borrowed from linear algebra, putting them in quotation marks.

IMPORTANT NOTE. One of the main features distinguishing the "linear" space $\mathscr{R}(X)$ from a true linear space is the possibility of "summation" of more than a countable number of terms (numbers, functions, etc.): for any uniformly bounded family $f_\alpha \in \mathscr{R}(X)$ we may put

$$\left(\bigoplus_\alpha f_\alpha \right)(x) = \sup_\alpha f_\alpha(x).$$

The above remark means that the idempotent analog to the space of integrable functions is simply the space of functions on X bounded from above. We denote it by $M(X)$: if X is compact, then

$$\begin{aligned} \mathscr{R}(X) = \{f : X \to \mathscr{R}\} \supset M(X) = \{f : \sup f < \infty\} \\ \supset C(X) = \{f : f \text{ is continuous on } X \text{ and nowhere equal to } -\infty\}. \end{aligned}$$

3.2. Operators. The main reason why the algebraic structures appear in the exposition is that the expression occurring on the left-hand side of the Bellman equation may be regarded as the value of a certain operator on the given function.

DEFINITION 3.1. Let $b \colon X \times X \to \mathscr{R}$ be a semicontinuous function (i.e., a function with a closed hypograph). Assume that the set X is compact. The

Bellman operator with kernel b is the operator $B: C(X) \to \mathscr{R}(X)$, defined by the formula

$$\forall f \in C(X), \quad (Bf)(x) = \max_{y \in X}\{b(x, y) + f(y)\}. \tag{3.1}$$

IMPORTANT NOTE. Actually, this definition is meaningful, provided that f is only semicontinuous. Moreover, the applicability of this definition may be significantly extended, if we replace max by sup in the above formulation, allowing thus for all $b \in \mathscr{R}(X \times X)$, $f \in \mathscr{R}(X)$ bounded from above. We shall denote this extension by the same symbol B.

PROPOSITION 3.1. *The Bellman operator is "linear":*

$$\forall f, g \in \mathscr{R}(X), \ \lambda \in \mathscr{R}, \qquad B(f \oplus g) = Bf \oplus Bg, \quad B(\lambda \odot f) = \lambda \odot Bf. \tag{3.2}$$

The property (3.2) is in fact characteristic of Bellman operators. If a certain continuity property holds for an operator B satisfying (3.2), than it can be represented in the form (3.1) with an appropriate function $b \in \mathscr{R}(X \times X)$ [9, 10]. Another condition sufficient for the existence of such a representation is "continual linearity", which means that (3.2) holds not only for two terms, but for any family of terms:

$$B\left(\bigoplus_{\alpha} f_\alpha\right) = \bigoplus B f_\alpha \tag{3.3}$$

provided that this family is "summable" (= uniformly bounded). In this case the kernel function $b(x, y)$ can be reconstructed from the values taken by B on the δ-functions

$$\delta_a(x) = \begin{cases} 0 & \text{if } x = a, \\ -\infty & \text{otherwise.} \end{cases}$$

Indeed, $b(x, y) = (B\delta_y)(x)$.

REMARK. Explicit conditions on the kernel b such that the associated Bellman operator maps the space $C(X)$ into itself can be formulated. This is of no special interest to us; see [10].

3.3. The algebra of operators. The general principle is to consider an algebraic object (the Bellman operator in our case) together with all objects possessing the same algebraic properties. So we introduce the set $\operatorname{End}\mathscr{R}(X)$ of all "linear" operators on $\mathscr{R}(X)$ and the subset $\operatorname{End} C(X)$ of those preserving the continuity of functions.

Operators from $\operatorname{End}\mathscr{R}(X)$ can be "added" and "multiplied" by scalars from $\mathscr{R}$. Clearly, these two operations preserve properties (3.2) and (3.3). For any two operators $B_1, B_2 \in \operatorname{End}\mathscr{R}(X)$, their composition $B_1 \circ B_2$ again belongs to $\operatorname{End}\mathscr{R}(X)$, the latter set being thus a semialgebra over the semiring $\mathscr{R}$.

The identity operator id belongs to $\operatorname{End}\mathscr{R}(X)$: its kernel is the function

$$\delta(x, y) = \begin{cases} 0 & \text{if } x = y, \\ -\infty & \text{otherwise.} \end{cases}$$

An operator $C \in \operatorname{End}\mathscr{R}(X)$ is called *invertible*, if there exists an endomorphism $C^{-1} \in \operatorname{End}\mathscr{R}(X)$ such that $C \circ C^{-1} = C^{-1} \circ C = \mathrm{id}$. The set of all invertible operators forms a group denoted $\operatorname{Inv}\mathscr{R}(X)$.

As in the case of linear operators, the two operators B, $\widetilde{B}$ are called *conjugate*, if there exists an invertible operator $C \in \operatorname{Inv}\mathscr{R}(X)$ such that

$$\widetilde{B} = C^{-1} \circ B \circ C. \tag{3.4}$$

This is an equivalence relation on the semialgebra of Bellman operators.

EXAMPLES. 1. For any continuous function $g: X \to \mathbb{R}$ ($g(x) \neq -\infty$) the operator

$$D_g: f \mapsto f + g \iff (D_g f)(x) = g(x) \odot f(x)$$

is the invertible "diagonal" operator with inverse D_{-g}.

2. For any one-to-one map $H: X \to X$ the operator $P_H: f \mapsto f \circ H$ is invertible with inverse $P_{H^{-1}}$. This operator and the one in the previous example belong to $\operatorname{End} C(X)$ if g is continuous and H is a homeomorphism.

3. If $C = D_g$, and B is a Bellman operator with the kernel b, then the conjugate operator $\widetilde{B} = C^{-1} \circ B \circ C$ has the kernel

$$\widetilde{b}(x, y) = -g(x) + b(x, y) + g(y),$$

which occurred before in our discussion of the notion of equivalent optimization problems in §2. The same refers to the second of the above examples.

DEFINITION 3.2. Two discrete time optimization problems with transition functions $b(x, y)$ and $\widetilde{b}(x, y)$ are called *equivalent* if the associated Bellman operators B and $\widetilde{B}$ are conjugate up to "multiplication" by a scalar: $\exists C \in \operatorname{Inv}\mathscr{R}(X)$, $\lambda \in \mathscr{R}$: $\widetilde{B} = \lambda \odot C^{-1} \circ B \circ C$.

Now we can explain why only the conjugation by matrices of the form D_g and P_H can be considered.

PROPOSITION 3.2 (V. N. Kolokoltsov, 1987). *An operator $C \in \operatorname{End}\mathscr{R}(X)$ is invertible if and only if it can be represented in the form $D_g \circ P_H$ for some everywhere finite function $g \in \mathscr{R}(X)$ and a certain one-to-one mapping H.*

Actually, a Bellman operator C is invertible if and only if its kernel $c(\cdot, \cdot)$ satisfies the condition $\forall x$, $\exists! y$: $c(x, y) \neq \infty$. Therefore the one-to-one mapping $H: x \mapsto y$, which will be denoted by T_C, appears. The optimization problem associated with an invertible operator is nothing more than a deterministic discrete time dynamical system on the phase space X, generated by the transformations H^t, $t \in \mathbb{Z}$, $H = T_C$.

REMARK. In the above exposition, the space $\mathscr{R}(X)$ can always be replaced by $M(X)$, provided that the kernel function is bounded from above.

IMPORTANT NOTE. The point is that although the group $\operatorname{Inv} C(X)$ is relatively small, it nevertheless acts transitively on the space $C(X)$. Indeed, any function $f_1 \in C(X)$ can be taken into any other f_2 by the invertible operator D_g with $g = f_1 - f_2$. But this is not true if the functions take the value $-\infty$.

3.4. Topology. Up to now only algebraic properties were under consideration. Now we invoke topological aspects. Our attention will be concentrated mostly on the spaces $M(X)$ and $C(X)$ (recall that the former one plays the role of the Lebesgue L^1-space from ordinary functional analysis).

DEFINITION 3.3. The Bellman operator (3.1) with kernel $b \in M(X \times X)$ is called *compact*, if:

(1) the phase space X is a metric compactum, and
(2) the function b is continuous and never has the value $-\infty$.

To motivate this definition, we note the following. Any compact Bellman operator takes its values in the space of all continuous functions. Moreover, when considered as a nonlinear operator acting on the Banach space $C(X)$ endowed with the usual norm $\|f\| = \max_{x \in X} |f(x)|$, it takes any bounded subset of the latter space into a set with compact closure (precompact). The proof is based on the following two lemmas.

LEMMA 3.1 (on the uniform continuity of images). *The image of the set $M^*(X) = M(X) \setminus \{f \equiv -\infty\}$ under a compact Bellman operator B is uniformly continuous, i.e., for any $\varepsilon > 0$ there exists a $\delta > 0$ such that*

$$\forall f \in M^*(X),\ \forall x_1, x_2 \in X, \quad \rho(x_1, x_2) < \delta \implies |Bf(x_1) - Bf(x_2)| < \varepsilon,$$

where ρ is the metric on X.

PROOF. Being continuous on the compact set $X \times X$, the kernel is equicontinuous, i.e., for any $\varepsilon > 0$ there exists a $\delta > 0$ such that

$$\forall x_1, x_2, y \in X, \quad \rho(x_1, x_2) < \delta \implies |b(x_1, y) - b(x_2, y)| < \varepsilon.$$

Therefore for $\rho(x_1, x_2) < \delta$ one has

$$Bf(x_1) = \sup_{y \in X}\{b(x_1, y) + f(y)\} \leqslant \sup_{y \in X}\{b(x_2, y) + f(y) + \varepsilon\} = Bf(x_2) + \varepsilon,$$

and the inverse inequality also holds: $Bf(x_2) \leqslant Bf(x_1) + \varepsilon$, whence uniform continuity. □

COROLLARY 3.1. *A compact Bellman operator always has continuous values:*

$$B(M^*(X)) \subset C(X).$$

LEMMA 3.2 (on monotonicity). *Any Bellman operator B preserves the Pareto partial order on $M(X)$: if $f_1 \geqslant f_2$ on X, then $Bf_1 \geqslant Bf_2$.*

PROOF. The inequality $f_1 \geqslant f_2$ is equivalent to the identity $f_1 \oplus f_2 = f_1$. The Bellman operator is "linear", hence $Bf_1 \oplus Bf_2 = Bf_1$. □

COROLLARY 3.2. *Any Bellman operator is nonexpanding in the Banach space* $C(X)$.

PROOF. If $\|f_1 - f_2\| \leqslant r$, then $f_1 \leqslant f_2 + r$, $f_2 \leqslant f_1 + r$. Applying Lemma 2 and using the "linearity" of B, we get $Bf_1 \leqslant Bf_2 + r$ and vice versa, whence we obtain the inequality $\|Bf_1 - Bf_2\| \leqslant r$. □

REMARK. Lemma 2 and Corollary 2 remain valid also for noncompact operators: in this case the corollary asserts that

$$\sup_X |f_1 - f_2| \leqslant r \implies \sup_X |Bf_1 - Bf_2| \leqslant r.$$

THEOREM 3.1. *A compact Bellman operator is continuous on the Banach space* $C(X)$ *endowed with the norm* $\|\cdot\|$ *and takes any bounded subset of* $C(X)$ *into a precompact one, thus being a compact operator in the usual sense of this term.*

PROOF. The continuity of B is proved in Corollary 2. For any bounded subset $U \subset C(X)$ the set $B(U)$ is uniformly continuous by Lemma 1 and is uniformly bounded by the constant

$$\max_{x, y \in X} |b(x, y)| + \sup_U \|f\|.$$

The Ascoli-Arzelà criterion of precompactness in $C(X)$ is applicable, hence $B(U)$ is precompact. □

3.5. Summary. We have introduced the notion of Bellman operator, the most important one for us, and analyzed the role of the continuity of the kernel in terms of nonlinear functional analysis.

§4. "Eigenfunctions"

4.1. "Spectral" equation. As the main result, the heuristic arguments listed in §2 lead to investigation of the *Bellman equation* (3.1) which, using the notation introduced in §3, can be written in a form quite similar to the spectral equation from linear algebra:

$$Bf = \lambda \odot f, \qquad \lambda \in \mathbb{R}, \; f \in M(X). \tag{4.1}$$

4.2. Quotient "projective" space. The Banach space $C(X)$ contains the 1-dimensional subspace of constants denoted by $\mathbb{R}$. The quotient space $C_\diamond(X) = C(X)/\mathbb{R}$ inherits the structure of a Banach space endowed with the quotient norm $\|f(\cdot)\|_\diamond = \min_{\lambda \in \mathbb{R}} \|\lambda + f\|$. The equivalence class of an element f will be denoted by $f + \mathbb{R}$.

The quotient space $C_\diamond(X)$ is obtained from $C(X)$ by "projectivization": elements f and $\lambda \odot f$ are identified if λ is "nonzero" (i.e., $\lambda \neq -\infty$).

The Ascoli-Arzelà criterion of precompactness in $C(X)$ implies the following precompactness criterion for $C_\diamond(X)$:

LEMMA 4.1. *Suppose that* X *is a compact metric space. Then a family*

$$\{f_\alpha + \mathbb{R} : f_\alpha \in C(X),\ \alpha \in \mathfrak{A}\}$$

is precompact in $C_\diamond(X)$, *if the family* $\{f_\alpha\}$ *is equicontinuous.*

PROOF. If the family $\{f_\alpha\}$ is equicontinuous, and X is compact, then the total oscillation $\sup_X f_\alpha - \inf_X f_\alpha$ is uniformly bounded over all $\alpha \in \mathfrak{A}$. Therefore one may add a certain constant λ_α (say, $\lambda_\alpha = -\inf_X f_\alpha$) to each f_α so that the new family $\{\lambda_\alpha + f_\alpha\}$ be uniformly bounded, therefore precompact by the Ascoli-Arzelà theorem. This immediately implies the precompactness of $\{f_\alpha + \mathbb{R}\}$ in $C_\diamond(X)$. □

Since any Bellman operator B commutes with the "multiplication" by scalars, the quotient operator $B_\diamond$ can be defined:

$$B_\diamond(f + \mathbb{R}) = Bf + \mathbb{R}, \tag{4.2}$$

which is a nonlinear operator. If B is compact, then by Lemma 1, §3, the quotient operator $B_\diamond$ is a compact operator from the space $C_\diamond(X)$ into itself. Moreover, the image $B_\diamond(C_\diamond(X))$ is precompact in $C_\diamond(X)$.

4.3. Existence of "eigenfunctions" for a compact Bellman operator. Now we can prove solvability of the "spectral" equation (4.1).

THEOREM 4.1. *Any compact Bellman operator possesses at least one continuous "eigenfunction".*

PROOF. One has only to apply the Schauder fixed point principle to $B_\diamond$: this is a compact operator taking a sufficiently big ball in $C_\diamond(X)$ into itself. Thus there exists a fixed point $f_* + \mathbb{R} \in C_\diamond(X)$ for $B_\diamond$:

$$B_\diamond(f_* + \mathbb{R}) = f_* + \mathbb{R},$$

which means that f_* together with a certain $\lambda \in \mathscr{R}$ satisfies (4.1). The number λ is finite, and f_* is continuous by construction. □

4.4. "Adjoint" operators. To prove the uniqueness of the "eigenvalue" for a compact Bellman operator, we need a "scalar product" in $M(X)$.

DEFINITION 4.1. The *standard "scalar product"* in $M(X)$ is the "bilinear" functional

$$\langle f, g\rangle = \sup_{x\in X}(f(x) + g(x)) = \bigoplus_{x\in X} f(x) \odot g(x).$$

In the same manner as this is done in the ordinary linear case, we define the "adjoint" operator B^* to any Bellman operator B by means of the identity

$$\forall f, g \in C(X), \quad \langle Bf, g\rangle = \langle f, B^* g\rangle.$$

PROPOSITION 4.1. *The "adjoint" operator always exists. Its kernel* b^* *is given by the formula*

$$b^*(x, y) = b(y, x)$$

(*the transposed matrix!*). *The "adjoint" to a compact operator is also compact.*

4.5. Uniqueness of the "spectrum". We start studying the set of solutions to the "spectral" equation.

THEOREM 4.2. *The "eigenvalue" for any compact Bellman operator is unique.*

PROOF. Let λ be one of the "eigenvalues" existing by virtue of Theorem 1. Since the "adjoint" operator B^* is also compact, there is at least one "eigenvalue" λ^* for B^*. Denote the corresponding "eigenfunctions" by f and f^* respectively. Since both of them are finite, $\langle f, f^*\rangle > -\infty$. But by definition one has

$$\lambda \odot \langle f, f^*\rangle = \langle Bf, f^*\rangle = \langle f, B^* f^*\rangle = \lambda^* \odot \langle f, f^*\rangle,$$

therefore $\lambda = \lambda^*$. But this means that any element of the "spectrum" of B equals one of B^*, whence the uniqueness. □

4.6. Corollaries and examples. For convenience we introduce some notation. The unique "eigenvalue" for a compact operator B will be denoted by $\operatorname{spec}(B)$. The set of corresponding "eigenfunctions" (the "eigenspace") is denoted by $\operatorname{es}(B)$.

COROLLARY 4.1 (I. V. Romanovskiĭ [16], 1967). *Suppose that B is a compact operator. Then for any function $f \in C(X)$ the iterates $B^t f$, $t = 0, 1, 2, \dots$, grow as an arithmetic progression: there exists $\lambda \in \mathbb{R}$ such that*

$$\|B^t f - f - t\lambda\| = O(1) \quad \text{as } t \to \infty$$

(here the iterates B^t are defined as $B^t = B \circ B^{t-1}$).

PROOF. Take any $f_* \in \operatorname{es}(B)$. Then $B^t f_* = t\lambda$, where $\lambda = \operatorname{spec}(B)$. Next, since all the B^t are nonexpanding,

$$\|B^t f - B^t f_*\| \leqslant \|f - f_*\| = O(1),$$

which immediately yields the required estimate.

COROLLARY 4.2 (A. Leizarowitz [32], 1985). *Any transition function continuous on the compact phase state X can be represented in the form*

$$b(x, y) = \lambda + g(y) - g(x) + b'(x, y),$$

where $b'(\cdot, \cdot)$ is a good nonpositive function, and g is continuous on X.

The problem of finding a "useful representation" (the term comes from A. Leizarowitz) was reduced to the Bellman equation in §2. In fact, an operator with a good nonpositive kernel function is completely characterized by the following property: the identical zero function 0 is the "eigenfunction" for the associated Bellman operator. Moreover, the reduction to good nonpositive form is exactly the transformation $B \mapsto B' = D_f^{-1} \circ B \circ D_f$, where $f \in \operatorname{es}(B)$. Indeed, $D_f(0) = f$, so $B'(0) = \lambda + 0$.

EXAMPLE 1. The spectrum may be nonunique if any of the two conditions of compactness of Bellman operators is violated. Indeed, if $B = D_g$ is the "diagonal" operator with the nonconstant function g, then (independently of the compactness of X) the "spectrum" contains at least all the values of g: the delta-function $\delta_a(x)$ satisfies (4.1) with $\lambda = g(a)$. This nonuniqueness of the spectrum is due to the discontinuity of the kernel $b(x, y) = \delta_y(x) \odot g(x)$.

EXAMPLE 2. If the other condition of compactness is violated, the spectrum may be nonunique as well. Let $X = \mathbb{R}^n$ be the noncompact phase space and assume that $b(x, y) = \Phi(y - x)$, where $F: \mathbb{R}^n \to \mathscr{R}$ is a concave function (as smooth as one wishes). In this case any linear function $f(x) = \langle p, x \rangle$, $p \in \mathbb{R}^{n*}$ satisfies (4.1):

$$\sup_{y \in \mathbb{R}^n} \{\Phi(y - x) + \langle p, y \rangle\} = \sup_{v \in \mathbb{R}^n} \{\Phi(v) + \langle p, x \rangle + \langle p, v \rangle\} = \langle p, x \rangle + \Phi^*(p),$$

where $\Phi^*(\cdot)$ is the Legendre transform for $-\Phi$. If Φ is not a linear function, then the set of points at which Φ^* is finite, consists of more than one point, hence the spectrum is "multiple".

PROPOSITION 4.2. *The "eigenspace" consisting of "eigenfunctions" with the same "eigenvalue", is indeed a "subspace"* (= *subsemimodule of* $M(X)$):

$$\mathrm{es}(B) \oplus \mathrm{es}(B) = \mathrm{es}(B), \qquad \mathbb{R} \odot \mathrm{es}(B) = \mathrm{es}(B).$$

This means that together with $f \in \mathrm{es}(B)$ any function $\lambda \odot f = \lambda + f$ is also an "eigenfunction". Therefore when analyzing the nonuniqueness of "eigenfunctions", we shall mean by this that $\mathrm{es}(B)$ cannot be generated by any single function.

EXAMPLE 3. Even if an operator is compact, its spectrum then being a singleton, the "eigenfunctions" can be nevertheless multiple in the above sense. The simplest possible case is the two-point set $X = \{1, 2\}$ with the following transition matrix $A = \|a_{ij}\|$, $a_{ij} = b(i, j)$, $i, j = 1, 2$:

$$\begin{pmatrix} 0 & -1 \\ -2 & 0 \end{pmatrix}.$$

Indeed, in this case we have a compact operator (since all the a_{ij} are finite), and both functions

$$f_1(1) = 0, \ f_1(2) = -2 \quad \text{and} \quad f_2(1) = -1, \ f_2(2) = 0$$

belong to $\mathrm{es}(B)$ with $\mathrm{spec}(B) = 0$, but $f_2 \neq \lambda \odot f_1$ for any $\lambda \in \mathscr{R}$.

The matrix case (that of a discrete compact X) was extensively analyzed by I. V. Romanovskiĭ [16], where he has given a complete description of "eigenfunction" in terms of maximal loops of the weighted graph representing the corresponding discrete optimization problem.

4.7. The vanishing viscosity technique. There is a transparent analogy between the spectral theory of Bellman operators and that of positive linear operators. For example, the method used to prove uniqueness of the spectrum is exactly the same as the one used in the demonstration of the Frobenius-Perron theorem on nonnegative matrices.

This analogy can be explained as follows. Let dx be any finite nonnegative measure on X such that open subsets have positive measure (such measure exists because X was assumed separable, hence it has a countable dense subset; the measure can be constructing by placing a mass of 2^{-k} at the kth point of this subset).

Then for any $f \in C(X)$ one has

$$\lim_{h \to 0^+} h \log \int_X \exp(f(x)/h)\, dx = \max_{x \in X} f(x).$$

Consider the family of operators B_h approximating a compact Bellman operator B with continuous kernel $b(\cdot\,,\cdot)$:

$$(B_h f)(x) = h \log \int_X \exp(h^{-1}(b(x\,,y) + f(y)))\, dy.$$

Then B_h is the composition $L_h \circ A_h \circ L_h^{-1}$, where $L_h f = h \log f$, and A_h is the linear (without quotation marks!) integral operator on $C(X)$ with the continuous everywhere positive kernel $a_h(x\,,y) = \exp(h^{-1} b(x\,,y))$. Applying the Frobenius-Perron theorem in its infinite-dimensional version [12], we conclude that there exists an eigenfunction $\varphi_h \in C(X)$ for A_h; therefore $f_h = L_h \Phi_h$ is an "eigenfunction" for B_h.

One can easily check that it is possible to pass to the limit as $h \to 0^+$ along some sequence $h_k \to 0^+$ in such a way that f_{h_k} is uniformly convergent in $C(X)$. Since $B_h \to B$, the limit function satisfies (4.1). □

Actually, the substitution $u \mapsto h^{-1} \log u$ of the dependent variable is used to transform the nonlinear Bürgers equation

$$\frac{\partial u}{\partial t} = \left(\frac{\partial u}{\partial x}\right)^2 + h \frac{\partial^2 u}{\partial x^2}$$

into the linear parabolic heat conduction equation

$$\frac{\partial u}{\partial t} = h \frac{\partial^2 u}{\partial x^2},$$

which can be explicitly investigated. The second order term in the nonlinear equation is a small viscosity added to the Hamilton-Jacobi equation. Since the latter is evidently related to optimization, there is no reason to wonder about the applicability of such an approach in some other situations; see [9, 14].

§5. Infinite extremals

In this section we give the definition of infinite extremals for the discrete optimization problem with transition function $b\colon X \times X \to \mathscr{R}$.

5.1. Definition of infinite extremals. Within the framework of our approach, we associate the notion of an infinite extremal with the Bellman operator B itself rather than with its kernel (the transition function). The properties of infinite extremals established in subsequent sections will motivate the choice of the suggested definition. The key point is formula (2.6) generating finite horizon extremals via the dynamic programming procedure.

MAIN DEFINITION 5.1. Let $B \in \operatorname{End} C(X)$ be the Bellman operator and $\widehat{f} \in \operatorname{es}(B)$ an "eigenfunction". An infinite trajectory $\mathbf{x} = \{x_t\}_{t=0}^{\infty}$ will be called *$\widehat{f}$-extremal*, if and only if

$$\forall t = 1, 2, \dots, \quad x_t \in \operatorname{Arg\,max}\{b(x_{t-1}, y) + \widehat{f}(y)\} \tag{5.1}$$

or, equivalently,

$$\forall t = 1, 2, \dots, \quad b(x_{t-1}, x_t) = \widehat{f}(x_{t-1}) - \widehat{f}(x_t) + \operatorname{spec}(B). \tag{5.2}$$

The set of all $\widehat{f}$-extremals will be denoted by $\operatorname{extr}_\infty(\widehat{f}, B)$. We shall say that $\mathbf{x}$ is an infinite extremal without referring to the choice of $\widehat{f}$, if it is an $\widehat{f}$-extremal for some $\widehat{f} \in \operatorname{es}(B)$; we shall denote

$$\operatorname{Extr}_\infty(B) = \bigcup_{\widehat{f} \in \operatorname{es}(B)} \operatorname{extr}_\infty(\widehat{f}, B).$$

Clearly, $\operatorname{Extr}_\infty(\lambda \odot C^{-1} \circ B \circ C) = T_C(\operatorname{Extr}_\infty(B))$ for any $C \in \operatorname{Inv} C(X)$, which means that the notion of the infinite extremal is invariant under inner automorphisms of the operator semialgebra.

Theorem 4.1 and the solvability of the recursive equation (5.1) imply the following statement.

THEOREM 5.1. *Let $B \in \operatorname{End} C(X)$ be a compact Bellman operator with continuous kernel, and $x_0 \in X$ an arbitrary initial state. Then there exists at least one infinite extremal $\mathbf{x}$ starting at the point x_0.*

Now we list some evident properties of infinite extremals defined in this way.

5.2. Generalized Euler-Lagrange condition.

DEFINITION 5.2. We shall say that a trajectory $\mathbf{x}$ (either finite or infinite) satisfies the *generalized Euler-Lagrange condition*, if for any two numbers $t_1 < t_2$ and any sequence of states $\{y_t : t_1 \leqslant t \leqslant t_2\}$ with $x_{t_i} = y_{t_i}$, $i = 1, 2$, we have the inequality

$$\sum_{t=t_1}^{t_2-1} b(x_t, x_{t+1}) \geqslant \sum_{t=t_1}^{t_2-1} b(y_t, y_{t+1}).$$

In other words, this means that the trajectory cannot be improved (with respect to the given criterion) by changing any finite number of its states.

In particular, if the trajectory $\mathbf{x}$ satisfies the generalized Euler-Lagrange condition, then for any $t > 0$ (and $t < N$ in the finite horizon case)

$$x_t \in \operatorname*{Arg\,max}_{y \in X} b(x_{t-1}, y) + b(y, x_{t+1})$$

which in the case of smooth transition function implies

$$(\partial b/\partial y)(x_{t-1}, x_t) + (\partial b/\partial x)(x_t, x_{t+1}) = 0,$$

that is, the usual discrete time analog of the Euler-Lagrange necessary conditions of optimality.

PROPOSITION 5.1. *Any infinite extremal satisfies the generalized Euler-Lagrange condition.*

The proof is a straightforward application of the Bellman optimality principle. □

5.3. First versus second order. As it follows from the previous subsection, the Euler-Lagrange condition reduces to something like a second-order difference equation. Therefore to "determine" such a trajectory, one needs either two initial states, or the initial and the final state.

For any fixed $\widehat{f} \in \mathrm{es}(B)$ the $\widehat{f}$-extremals are "determined" only by the initial condition, as can be seen from (5.1). This property can be formulated as follows.

PROPOSITION 5.2. *Let* $\mathbf{x}$ *be any* $\widehat{f}$*-extremal, and* $\mathbf{x}'$ *be another* $\widehat{f}$*-extremal with the same "eigenfunction"* $\widehat{f}$ *such that* $x'_0 = x_N$ *for some* $N \in \mathbb{N}$. *Then the trajectory*

$$\widehat{x}_t = \begin{cases} x_t & \text{for } t = 0, 1, \dots, N, \\ x'_{t-N} & \text{for } t = N+1, N+2, \dots, \end{cases}$$

is also $\widehat{f}$*-extremal.*

In other words, infinite extremals associated with the same "eigenfunction" can be pasted together.

EXAMPLE 5.1. This is not true for different "eigenfunctions". The simplest example is the case of the translation invariant transition function $b(x, y) = \Phi(y - x)$ on $X = \mathbb{R}^n$, though this is in the noncompact case.

For the associated Bellman operator, all the linear functions belong to $\mathrm{es}_\lambda(B)$ provided that λ is a finite value of the Legendre transform of Φ (see Example 4.2).

The corresponding infinite extremals are straight lines (or, more exactly, vector-valued arithmetic progressions). In general, the vectors generating these progressions can be different, so when pasted together, segments of different extremals from the class $\mathrm{Extr}_\infty(B)$ can form "broken lines" which do not satisfy even the generalized Euler-Lagrange conditions.

5.4. Good trajectories. There is a useful weakening of the notion of infinite extremal which does not involve any reference to a specified "eigenfunction". This notion was also suggested by Gale [30].

DEFINITION 5.3. Let B be a compact Bellman operator with $\operatorname{spec}(B) = \lambda$. An infinite trajectory $\mathbf{x}$ is called *good*, if

$$\left|\sum_{t=0}^{N-1} b(x_t, x_{t+1}) - N\lambda\right| = O(1).$$

The following properties of good trajectories are evident.

PROPOSITION 5.3. 1. *Any $\widehat{f}$-extremal is a good trajectory. Therefore good trajectories always exist.*

2. *Any trajectory differing from a good one by a finite number of states, is also good.*

3. *The notion of a good trajectory is invariant under "conjugation": if $\mathbf{x}$ is a good trajectory for $B \in \operatorname{End} C(X)$, and $B' = \lambda \odot C^{-1} \circ B \circ C$, then $\mathbf{x}' = T_C(\mathbf{x})$ is good for B'.*

5.5. Examples. Here are some examples of trajectories from the class $\operatorname{extr}_\infty(B)$ which demonstrate some important features.

EXAMPLE 5.1. Let $X = \{1, 2, 3\}$ be the three point phase state with the transition matrix

$$\begin{pmatrix} -100 & -1 & -100 \\ 1 & -100 & -100 \\ \alpha & \beta & -100 \end{pmatrix}, \qquad \alpha, \beta < 1.$$

Then it is more or less evident (since -100 is almost $-\infty$), that there are only two infinite trajectories starting at the point 3 and satisfying the generalized Euler-Lagrange conditions. The associated cost flows are

$$\alpha, \alpha - 1, \alpha, \alpha - 1, \dots \quad \text{and} \quad \beta, \beta + 1, \beta, \beta + 1, \dots.$$

Both trajectories are good, and for certain α, β neither of them overtakes the other.

This problem possesses a unique "eigenfunction"

$$f(1) = 0, \quad f(2) = 1, \quad f(3) = \max(\alpha, \beta + 1).$$

Therefore for the case $\alpha > \beta+1$ the first of the two trajectories is f-extremal, while the opposite inequality implies the optimality of the second one. This answer coincides with that given by the heuristic argument that the average cost flow must be greater for the optimal path: the average equals to $\alpha - 1/2$ and $b + 1/2$ respectively. Thus the heuristics is justified.

EXAMPLE 5.2. Now we consider the 5-point phase state with transition matrix

$$\begin{pmatrix} * & * & * & * & * \\ 10 & 0 & * & * & * \\ * & 1 & * & 2 & * \\ * & * & * & 0 & 5 \\ * & * & * & * & * \end{pmatrix},$$

where stars stand for sufficiently large negative values (almost $-\infty$).

In this case there are also two infinite trajectories starting at the point 3 and satisfying the generalized Euler-Lagrange condition: $3, 2, 2, 2, \ldots$ and $3, 4, 4, 4, \ldots$; their cost flows are $1, 1, 1, \ldots$ and $2, 2, 2, \ldots$ respectively, and from a superficial point of view it is evident that the second one is preferable.

But if you look for finite horizon optimal paths starting at the center point, you will find that there are only two of them satisfying the generalized Euler-Lagrange condition, namely, $3, 2, 2, \ldots, 2, 1$ and $3, 4, 4, \ldots, 4, 5$, with the former yielding the greater value of the integral cost (11 versus 7). Therefore the finite horizon approximation leads to a very different answer.

This happens because the associated Bellman operator possesses *two* independent "eigenfunctions":

$$f_-(1) = -10, \; f_-(2) = 0, \; f_-(3) = 1, \; f_-(4) = *, \; f_-(5) = *,$$
$$f_+(1) = *, \; f_+(2) = *, \; f_+(3) = 2, \; f_+(4) = 0, \; f_+(5) = -5.$$

The role of "eigenfunctions" may consist in compensating possible disadvantages of the last step, so only the "recurring" states are to be taken into account (for more details see the next section). In our case the two trajectories correspond to the two "eigenfunctions", and there is no way to decide between them without invoking some additional arguments.

5.6. Conclusions. The notion of infinite extremal generated by a certain "eigenfunction" of the Bellman operator leads to an object which possesses all the intuitive apriori properties, including that of invariance. The only problem is how to choose between different solutions of the Bellman equation. Without any additional information about the optimization problem, one cannot decide about the "proper" function, since the symmetry group of invertible Bellman transformations acts transitively on the space $C(X)$.

§6. "Projectors" onto the "eigenspace"

6.1. Extended data for an optimization problem: the terminant together with the transition function. In this section we discuss ways of passing to the limit in the finite horizon terminal optimization problem in order to obtain an appropriate definition of infinite extremals.

From §4 we may easily conclude that without loss of generality any compact Bellman operator may be regarded as normalized to satisfy the condition

$$\operatorname{spec}(B) = 0. \tag{6.1}$$

For this we must replace the initial operator B by the new one $\lambda \odot B$, where $\lambda = -\operatorname{spec}(B)$. It follows from the above exposition that such a normalization does not affect any properties of the infinite extremals.

For an operator B satisfying (6.1) the *orbit* $\{B^n f\}_{n=0}^{\infty}$ of any function $f \in C(X)$ is a bounded equicontinuous (uniformly continuous and bounded) family in $C(X)$, therefore its closure is compact. (The same holds for any function $f \in \mathscr{R}(X)$, $f \not\equiv -\infty$ starting from $n = 1$.)

A passage to the limit as $N \to \infty$ in the problem

$$\sum_{t=0}^{N-1} b(x_t, x_{t+1}) + f(x_N) \to \max, \qquad x_0 = a, \tag{6.2}$$

within the general spirit of the suggested approach means that we must find a function $\widehat{f} \in C(X)$ which is an "eigenfunction", i.e., $f \in \mathrm{es}(B)$, and which inherits certain properties of the terminant f in (6.2). Since the operator B is assumed to be fixed, the correspondence $f \mapsto \widehat{f}$ may be written in operator form

$$f \mapsto \widehat{f} = \Omega_B f; \tag{6.3}$$

sometimes the explicit dependence on B will be omitted.

The correspondence $f \mapsto \Omega f$ must satisfy the following natural list of conditions:

(1) Ω_B is a "linear" operator: $\forall B$, $\Omega_B \in \operatorname{End} C(X)$;

(2) Ω_B is invariantly associated with B: if $B = \lambda \odot C^{-1} \circ B' \circ C$, $C \in \operatorname{Inv} C(X)$, $\lambda \in \mathbb{R}$, then

$$\Omega_B = C^{-1} \circ \Omega_{B'} \circ C; \tag{6.4}$$

(3) Ω_B is the identity on the "eigenspace" of B;

(4) for any $f \in C(X)$ the image $\Omega_B f$ is an "eigenfunction" of B.

The last two conditions together mean that Ω_B is an idempotent projector from $C(X)$ onto $\mathrm{es}(B) \subset C(X)$.

A priori it is not clear whether such an operator exists or not. In fact, we prove below that for any compact operator B there always exists a projector Ω_B satisfying all the above properties. Moreover, an axiomatic condition whose additon to the above list implies uniqueness of the projector can be formulated. But we would like to start with an important example, proving at the same time the existence assertion on the projectors. From now on we assume that $\operatorname{spec} B = 0$.

6.2. The McKenzie projector. Let $\mathbf{x} = \{x_t\}_{t=0}^{\infty}$ be an arbitrary infinite good trajectory, and $a \in X$ an arbitrary initial state.

Define the value

$$\gamma(a, f, \mathbf{x}) = \begin{cases} \limsup\limits_{N\to\infty} \left\{ \sum_{t=0}^{N-1} b(x_t, x_{t+1}) + f(x_N) \right\} & \text{if } x_0 = a, \\ -\infty & \text{otherwise.} \end{cases}$$

From the definition of a good trajectory it follows that $\limsup$ exists.

Next, define the function

$$\widehat{f}\colon a \mapsto \widehat{f}(a) = \sup_{\mathbf{x}} \{\gamma(a, f, \mathbf{x}) : \mathbf{x} \text{ is a good trajectory}\}.$$

The operator $\Gamma = \Gamma_B \colon f \to \widehat{f}$ satisfies the following conditions.

PROPOSITION 6.1. *Γ_B is a "linear" operator, and*

$$B\Gamma_B = \Gamma_B,$$

so Γ_B is the projector onto the "eigenspace" of B. The correspondence $B \mapsto \Gamma_B$ is invariant under conjugations: if $B = C^{-1} \circ B' \circ C$, $C \in \operatorname{Inv} C(X)$, then $\Gamma_B = C^{-1} \circ \Gamma_{B'} \circ C$.

PROOF. The "linearity" of the operator Γ_B follows immediately from definitions. For a fixed $a \in X$, $\mathbf{x} \in X^\infty$, the value γ is "linear" in f, as is implied by the identity

$$\limsup_{k\to\infty} \max(\alpha_k, \beta_k) = \max\left(\limsup_{k\to\infty} \alpha_k, \limsup_{k\to\infty} \beta_k\right),$$

valid for all pairs of real sequences. Since $\limsup_{k\to\infty} \alpha_k = \lim_{k\to\infty} \bigoplus_{j=k}^\infty \alpha_j$, this is a consequence of the identity

$$\lim_k \bigoplus_{j=k}^{\infty} (\alpha_j \oplus \beta_j) = \left(\lim_k \bigoplus_{j=k}^{\infty} \alpha_j\right) \oplus \left(\lim_k \bigoplus_{j=k}^{\infty} \beta_j\right).$$

Taking the supremum over all $\mathbf{x}$'s does not affect this "linearity".

A straightforward argument proves that for $\widehat{f} = \Gamma f$ one has $B\widehat{f} = \widehat{f}$. Indeed, if $\mathbf{x} = \{x_t\}$ and $\mathbf{y} = \{y_t\}$, $y_t = x_{t+1}$, then by the definition of γ

$$\gamma(x_0, f, \mathbf{x}) = b(x_0, x_1) + \gamma(x_1, f, \mathbf{y}).$$

Taking the supremum on both sides of this equality, one gets

$$\begin{aligned} \Gamma_B f(x_0) &= \sup_{\mathbf{x}} \gamma(x_0, f, \mathbf{x}) = \sup_{x_1, \mathbf{y}} \{b(x_0, x_1) + \gamma(x_1, f, \mathbf{y})\} \\ &= \sup_{x_1} (b(x_0, x_1) + \Gamma_B(x_1)) = B\Gamma_B f(x_0). \end{aligned}$$

The invariance comes directly from the definitions. □

The projector Γ_B constructed this way plays an essential role in the classical definition of infinite extremals. We shall call Γ_B the *McKenzie projector*.

NOTE 6.1. The two conditions

$$\Omega_B \circ \Omega_B = \Omega_B \quad \text{(the idempotent property)}, \qquad B \circ \Omega_B = \Omega_B,$$

taken together, mean that Ω_B is the projector onto the space of fixed points of B:

$$\forall f \in C(X), \quad \Omega_B f \in \operatorname{es}(B), \qquad \forall f \in \operatorname{es}(B), \quad \Omega_B f = f.$$

Now we are proceeding with the main result of this section. To motivate it, we point out the following

PROPOSITION 6.2. *For any finite horizon optimization problem, we have*

$$\operatorname{extr}_N(B, f)\big|_{[0, N-1]} = \operatorname{extr}_{N-1}(B, Bf).$$

In other words, to solve the N-step problem with terminant f is equivalent to solving the $(N-1)$-step one with terminant Bf.

PROOF. This is the Bellman optimality principle *verbatim.* □

Therefore the projector Ω_B associating an "eigenfunction" with any terminant f must satisfy the additional condition

$$\Omega_B \circ B = \Omega_B \tag{6.5}$$

if we want it to preserve the Bellman principle. Note that the latter condition means that Ω_B takes a constant value on any orbit of the operator B.

THEOREM 6.1. *Let B be a Bellman operator with continuous kernel on a compact space X, normalized by the condition* $\operatorname{spec} B = 0$. *Then there exists a unique projector $\Omega_B \colon C(X) \to \operatorname{es}(B)$ satisfying the above list of conditions and constant on the orbits of B in the sense of* (6.5).

NOTE 6.2. The algebraic conditions imposed on Ω_B in the nonnormalized case have the form

$$\Omega_B^2 = \Omega_B, \qquad \Omega_B \circ B = B \circ \Omega_B = \lambda \odot \Omega_B,$$

where $\lambda = \operatorname{spcc} B$.

PROOF. *Existence.* Let $\omega_N(f)$ be the closure of the family $\{B^n f\}_{n=N}^{\infty}$. We have mentioned above that $\omega_N(f)$ is compact in $C(X)$, and $\omega_1(f) \supseteq \omega_2(f) \supseteq \cdots$. So there is a nonempty intersection $\omega(f) = \bigcap_{N\geqslant 1} \omega_N(f)$.

Let Ωf be the "continual integral", or the "sum" of an uncountable number of functions

$$\Omega f = \bigoplus_{\varphi \in \omega(f)} \varphi \quad \text{or} \quad \Omega f(x) = \max_{\varphi(x)\,:\,\varphi\in\omega(f)} \varphi(x)$$

(which is the same in standard terms). Here max is correctly used, since $\omega(f)$ is compact in $C(X)$. We have the following formula for Ωf:

$$\Omega f = \lim_{N\to\infty} \bigoplus_{n=N}^{\infty} B^n f \tag{6.6}$$

or

$$\Omega = \lim_{N\to\infty} \bigoplus_{n=N}^{\infty} B^n. \tag{6.7}$$

The limit in (6.6) is taken in the norm topology on $C(X)$, and in (6.7) in the strong operator topology.

From (6.7) we deduce all desired properties. "Linearity" is evident; if $B = C^{-1} \circ B' \circ C$ and $\Omega' = \Omega_{B'}$, then

$$C^{-1} \circ \Omega' \circ C = \lim_{N\to\infty} C^{-1} \circ \left(\bigoplus_{n=N}^{\infty} B^n \right) \circ C = \lim_{N\to\infty} \bigoplus_{n=N}^{\infty} (C^{-1} \circ B \circ C)^n$$

whence comes the invariance.

If $Bf = f$ then $B^n f = f$ and $\Omega f = f$. Next,

$$\forall f \in C(X), \quad B\Omega f = B \lim_{N\to\infty} \bigoplus_{n=N}^{\infty} B^n f = \lim_{N\to\infty} \bigoplus_{n=N}^{\infty} B^{n+1} f = \Omega f$$

whence $\Omega f \in \mathrm{es}(B)$, that is, $\Omega(C(X)) = \mathrm{es}(B)$.

Finally, $\Omega B f = \lim_{N\to\infty} \bigoplus_{n=N}^{\infty} B^{n+1} f = \Omega f$, therefore (6.5) holds.

Uniqueness. Let $\widetilde{\Omega}$ satisfy all the conditions imposed on projectors together with (6.5). Then

$$\forall f \in C(X), \quad \Omega f \in \mathrm{es}(B),$$

so $\widetilde{\Omega} \circ \Omega f = \Omega f$. Therefore

$$\Omega f = \widetilde{\Omega}\Omega f = \widetilde{\Omega} \circ \left(\lim_{N\to\infty} \bigoplus_{n=N}^{\infty} B^n f \right) = \lim_{N\to\infty} \bigoplus_{n=N}^{\infty} \widetilde{\Omega} B^n f = \lim_{N\to\infty} \bigoplus_{n=N}^{\infty} \widetilde{\Omega} f = \widetilde{\Omega} f$$

(we again used the idempotency of "addition" $\oplus$). So the desired uniqueness is established.

NOTE 6.3. The construction of the projector Ω is very geometric. The set $\omega(f)$ is the ω-limit set of the orbit $\{B^n f\}_{n=0}^{\infty}$ as it is defined in the theory of dynamical systems, see [15]: this is the set of accumulation points of the orbit $B^n f$ in the functional space $C(X)$. By quite general reasons $\omega(f)$ in invariant by B: if $g \in C(X)$ is the uniform limit for a subsequence $B^{n_k} f$, then Bg is the one for $B^{n_k+1} f$. Moreover, $B\omega(f) = \omega(f)$: for any $g = \lim_{k\to\infty} B^{n_k} f$ there exists at least one accumulation point $g' \in \omega(f)$ for the subsequence $B^{n_k-1} f$ such that $Bg' = g$.

The operator B, being "linear" on any set $M \subset C(X)$, leaves its "barycenter" $\bigoplus_M \varphi = \bigoplus_{\varphi\in M} \varphi$ invariant:

$$B \bigoplus_{M} \varphi = \bigoplus_{M} B\varphi = \bigoplus_{B(M)} \varphi = \bigoplus_{M} \varphi.$$

Hence we obtain the fixed point of B in purely topological terms.

EXAMPLE 6.2. Consider the finite discrete compact set X consisting only of five points $X = \{1, 2, 3, 4, 5\}$. Consider the transition matrix from Example 5.2.

Let $f = \{f_i = f(i),\ i = 1, \dots, 5\}$ be the zero function: $f_i = 0\ \forall i$.

Then $\Gamma f(2) = \Gamma f(4) = 0$, $\Gamma f(3) = 2$ (other values are of no interest). Another simple calculation shows that $\Omega f(2) = 10$, $\Omega f(4) = 5$ and $\Omega f(3) = 11$. So we have $\Omega f \neq \Gamma f \bmod \mathbb{R}$, which implies that that $\Gamma B \neq \Gamma$ (see Theorem 6.1).

This fact results in what we have already seen: the unique Γf-extremal starting at 3 is $\{3, 2, 2, \dots\}$, the unique Ωf-extremal is $\{3, 4, 4, \dots\}$.

The importance of the projection Γ for the classical theory will be clear from §8 below.

§7. Cyclic optimization and the "trace" formula

In this short section we give an explicit representation for the "eigenvalue" of a compact Bellman operator in terms of maximal cycles.

7.1. Cyclic optimization. A natural way to get rid of the terminant in the criterion, as well as of the initial state, is to consider the cyclic optimization problem of the form

$$\sum_{t=0}^{N-1} b(x_t, x_{t+1}) \to \max, \qquad x_0 = x_N. \tag{7.1}$$

Denote the set of solutions to this problem by $\operatorname{extr}_N^{\circlearrowleft}(B) \subset X^N$. Evidently, this set is invariantly associated with B: if $B = \lambda \odot C^{-1} \circ B' \circ C$, then $\operatorname{extr}_N^{\circlearrowleft}(B) = T_C(\operatorname{extr}_N^{\circlearrowleft}(B'))$.

7.2. The "trace" of a Bellman operator. If $B \in \operatorname{End} C(X)$, then the value

$$\operatorname{tr} B = \sup_{x\in X} b(x, x) = \bigoplus_{x\in X} b(x, x) \tag{7.2}$$

is called its "*trace*". The "trace" satisfies the following properties.

1. It is "linear" and invariant with respect to inner automorphisms:

$$\operatorname{tr}(B_1 \oplus B_2) = \operatorname{tr} B_1 \oplus \operatorname{tr} B_2, \qquad \operatorname{tr}(\lambda \odot B) = \lambda \odot \operatorname{tr} B, \tag{7.3}$$

$$\operatorname{tr}(C^{-1} \circ B \circ C) = \operatorname{tr} B \tag{7.4}$$

for all $B_i \in \operatorname{End} C(X)$, $C \in \operatorname{Inv} C(X)$.

2. The value of problem (7.1) is equal to $\operatorname{tr}(B^N) = \operatorname{tr}(B \circ B \circ \cdots \circ B)$ (N times).

7.3. The "trace" formula. We begin by establishing a formula which relates the spectrum of a compact Bellman operator to the trace of the "sum of a geometric progression".

THEOREM 7.1. *Let B be a compact Bellman operator with the zero "eigenvalue"* $\operatorname{spec}(B) = 0$. *Then*

$$\operatorname{tr}\left(\bigoplus_{n=1}^{\infty} B^n\right) = 0. \tag{7.5}$$

PROOF. Due to the invariance and "linearity" of the trace, without loss of generality we may assume that the kernel b of the operator is good nonpositive. This implies immediately that along any cyclic orbit $x_0, x_1, \dots, x_{N-1}$, $x_N = x_0$ the sum $\sum_{t=0}^{N-1} b(x_t, x_{t+1})$ is nonpositive, therefore the left-hand side of (7.5) is also nonpositive.

Suppose that it is negative (less than a certain $-\varepsilon < 0$). Consider the multivalued map $x \mapsto F(x) = \{y \in X : b(x, y) = 0\} \subset X$. Since b is continuous and X compact, for any $\varepsilon > 0$ there is $\delta > 0$ such that $\operatorname{dist}(y, F(x)) < \delta$ implies $b(x, y) > -\varepsilon$ uniformly in (x, y).

Now take any infinite extremal, say generated by the identically zero "eigenfunction". Since X is compact, this extremal must have an accumulation point $x_* \in X$, which means that there will be a pair of states x_τ, $x_{\tau+k}$ at least $\delta/2$-close to x_*, hence δ-close to each other. In other words, the trajectory makes almost δ-closed loop.

By the choice of δ we have $0 = b(x_\tau, x_{\tau+1}) > b(x_{\tau+k}, x_{\tau+1}) > -\varepsilon$, therefore by δ-changing only one point of the trajectory, we get an exactly closed loop with ε-close value:

$$b(x_{\tau+k}, x_\tau) + \sum_{t=\tau+1}^{\tau+k-1} b(x_t, x_{t+1}) > -\varepsilon,$$

therefore $\operatorname{tr} B^k > -\varepsilon$, in contradiction with the negativity assumption. □

7.4. Corollaries.

COROLLARY 7.1 [16]. *The "eigenvalue" of a compact Bellman operator is given by the formula*

$$\operatorname{spec}(B) = \sup_{n \geqslant 1} n^{-1} \operatorname{tr} B^n$$

which is the maximal average value of all closed loops.

REMARK. In fact, this corollary is much weaker than the theorem itself: to prove the former, one has only to establish the estimate $|\operatorname{tr} B^n| = O(1)$ for the case $\operatorname{spec}(B) = 0$.

NOTE 7.1. If X is a discrete compact set consisting of $\#X$ points, formula (7.5) may be strengthened:

$$\operatorname{tr}\left(\bigoplus_{n=1}^{\#X} B^n\right) = 0. \tag{7.5'}$$

Indeed, the accumulation argument in this case is trivial: no later than after $N = \#X$ steps some of the states must be repeated in any infinite trajectory.

In fact, the maximal cycles (loops) in the discrete case correspond to "eigenfunctions": if $\mathbf{x}_\circlearrowleft$ is an N-periodic loop with value equal to zero (we again consider the case $\operatorname{spec}(B) = 0$), then the function

$$f(a) = \limsup_N \sup_{\substack{\mathbf{y} : y_0 = a, \\ y_N \in \mathbf{x}_\circlearrowleft}} \sum_{t=0}^{N-1} b(y_t, y_{t+1}) \tag{7.6}$$

is an "eigenfunction" with zero "eigenvalue", and the functions corresponding to disjoint maximal cycles are "linearly" independent. This problem was extensively treated in [16].

§8. A bridge to classical theory

8.1. Classical notions of optimality. We recall here the notions of optimality based on the overtaking criterion as developed by Gale and von Weizsacker.

Let B be a compact Bellman operator with transition function $b: X \times X \to \mathbb{R}$. To any trajectory $\mathbf{x} = \{x_t\}_{t=0}^{\infty}$ we associate the cost flow $\mathbf{c} = \mathfrak{C}(\mathbf{x}) = \{c_t\}_{t=1}^{\infty} \in \mathbb{R}^{\mathbb{N}}$ by the formula

$$c_N = \sum_{t=0}^{N-1} b(x_t, x_{t+1}).$$

DEFINITION 8.1. 1. A trajectory $\mathbf{x}$ *overtakes* another trajectory $\mathbf{x}'$ if both of them have the same initial point and the associated cost flows $\mathbf{c}$ and $\mathbf{c}'$ satisfy the condition

$$\liminf_{t\to\infty}(c_t - c_t') \geqslant 0. \tag{8.1}$$

We shall denote this fact by means of the inequality $\mathbf{x} \geqslant \mathbf{x}'$. Sometimes, when it is necessary, the notation will include the Bellman operator: $\mathbf{x} \overset{B}{\geqslant} \mathbf{x}'$.

If $\liminf_{t\to\infty}(c_t - c_t') \geqslant \varepsilon > 0$, then we shall say that $\mathbf{x}$ *supertakes* $\mathbf{x}'$.

2. A trajectory $\mathbf{x}$ is called *overtaking*, if it overtakes any other trajectory with the same starting point.

3. A trajectory is called *weakly optimal*, if it is not supertaken by any other trajectory.

The condition (8.1) means that for any positive ε the value of a finite N-segment of the path $\mathbf{x}$ is greater than that of $\mathbf{x}'$ up to ε for all N large enough:

$$\forall \varepsilon > 0, \exists N = N(\varepsilon): \quad c_t > c_t' - \varepsilon. \tag{8.2}$$

If in this definition ε can be zero, the notion of strict overtaking appears, but we will not dwell on these matters; see [1].

8.2. The McKenzie projector and the overtaking criterion. The McKenzie projector was defined in §6. We recall its definition in somewhat different terms. Again we consider the case of compact operators with $\operatorname{spec}(B) = \lambda$.

Let $\mathscr{B}_f$ be the functional defined on infinite trajectories by the formula

$$\mathscr{B}_f(\mathbf{x}) = \limsup_{N\in\mathbb{N}} \left\{ \sum_{t=0}^{N-1} b(x_t, x_{t+1}) - N\lambda + f(x_N) \right\}. \tag{8.3}$$

The McKenzie projector is the correspondence

$$\Gamma: f \mapsto \Gamma f, \qquad \Gamma f(a) = \sup_{\mathbf{x}\,:\, x_0 = a} \mathscr{B}_f(\mathbf{x}). \tag{8.4}$$

It was shown in §6 that Γ is indeed a "linear" operator which takes its values in the "eigenspace" $\operatorname{es}(B)$.

The most important role in the classical infinite horizon optimization is played by the "eigenfunction" $\widehat{f}_0 = \Gamma 0$ (the value taken by the McKenzie projector on the identically zero function).

PROPOSITION 8.1. *The function* $\widehat{f}_0$ *is continuous on* X *provided that* B *is a compact operator.*

Indeed, all the "eigenfunctions" of a compact operator are continuous. The main property of $\widehat{f}_0$ is given by the following proposition.

LEMMA 8.1. *If a trajectory* $\mathbf{y}$ *is not an* $\widehat{f}_0$*-extremal, then it does not overtake a certain trajectory* $\mathbf{x}^*$ *starting at the same point.*

PROOF. Without loss of generality we may assume that the condition

$$b(y_t, y_{t+1}) + \widehat{f}_0(y_{t+1}) = \widehat{f}_0(y_t) + \lambda \tag{8.5}$$

is violated from the very beginning (i.e., for $t = 0$), the difference between the left- and the right-hand side of (8.5) being a certain negative $-\varepsilon < 0$. For simplicity we subtract λ from the transition function, passing thus to the normalized case.

This means that for any finite $N \geqslant 1$ we have

$$\widehat{f}_0(y_0) - \varepsilon > \widehat{f}_0(y_N) + \sum_{t=0}^{N-1} b(y_t, y_{t+1}),$$

therefore by the definition of $\widehat{f}_0$

$$\begin{aligned}\limsup_{M\to\infty} \sum_{t=0}^{M} b(y_t, y_{t+1}) &= \sum_{t=0}^{N-1} b(y_t, y_{t+1}) + \limsup_{M\to\infty} \sum_{t=N}^{M} b(y_t, y_{t+1}) \\ &\leqslant \sum_{t=0}^{N-1} b(y_t, y_{t+1}) + \widehat{f}_0(y_N) < \widehat{f}_0(y_0) - \varepsilon,\end{aligned}$$

and on the other hand there exists a certain trajectory $\mathbf{x}^*$ such that

$$\limsup_{N\to\infty} \sum_{t=0}^{N-1} b(x_t^*, x_{t+1}^*) > \sup_{\mathbf{x}\,:\, x_0=y_0} \limsup_{N\to\infty} \sum_{t=0}^{N-1} b(x_t, x_{t+1}) - \varepsilon/2 = \widehat{f}_0(y_0) - \varepsilon/2.$$

Finally we get

$$\begin{aligned}&\limsup_{N\to\infty} \sum_{t=0}^{N} (b(x_t^*, x_{t+1}^*) - b(y_t, y_{t+1})) \\ &\qquad \geqslant \limsup_{N\to\infty} \sum_{t=0}^{N-1} b(x_t^*, x_{t+1}^*) - \limsup_{N\to\infty} \sum_{t=0}^{N-1} b(y_t, y_{t+1}) > \varepsilon - \varepsilon/2 > 0,\end{aligned}$$

which proves the claim, since for $\mathbf{y}$ to overtake $\mathbf{x}^*$ it is necessary that this upper limit be nonpositive. □

As a corollary to this assertion, we obtain the following result.

THEOREM 8.1. *If an overtaking path exists, then it must be the infinite* $\widehat{f}_0$*-extremal.*

It is very likely that all the weakly optimal paths, if any, also belong to $\mathrm{Extr}_\infty(B)$, that is, they are generated by certain "eigenfunctions".

8.3. Remarks and discussion. From the above exposition one can come to the following conclusions. First, the notion of overtaking *is not invariant under conjugation*. This means that the relation $\mathbf{x} \geqslant \mathbf{y}$ may be affected when passing to another transition function equivalent to the initial one. The McKenzie projector representation makes this point clear: the generating "eigenfunction" is the image under Γ of the identical zero function, although this function is not distinguished among the other functions.

In other words, to define a partial order of the above kind that would be invariant, one needs to fix a certain terminant function and proceed as in the case of the McKenzie projector.

Another remark concerns with the McKenzie projector itself. As was pointed out in §6, there exists a unique projector onto the 'eigenspace' that satisfies the condition $\Omega \circ B = \lambda \odot \Omega$. In general the McKenzie projector does not satisfy it (e.g. in the case of Examples 5.2, 6.2). This can be interpreted as a kind of noncommutativity of the two passages to the limit. When dealing with the infinite horizon optimization, one must pass to limit along an individual trajectory and take the supremum over all trajectories. Suppose that $\mathscr{B}^N(\mathbf{x})$ is a certain finite-horizon criterion, for example, $\sum_{t=0}^{N-1} b(x_t, x_{t+1})$. Then in general

$$\sup_{\mathbf{x}} \lim_{N\to\infty} \mathscr{B}^N(\mathbf{x}) \neq \lim_{N\to\infty} \sup_{\mathbf{x}} \mathscr{B}^N(\mathbf{x})$$

whatever this means.

Both arguments above speak in favor of the definition of infinite extremals as the trajectories of the class $\operatorname{extr}_\infty(B, \Omega f)$ provided that a certain a priori terminant f is given. But all the objections disappear if we consider the special case of *operators having a one-dimensional "eigenspace"*, that is, possessing a unique "eigenfunction" up to addition of a constant. In this case one can take this "eigenfunction" as the generator for extremals, this choice being invariant by conjugations. Moreover, in such a situation all projectors differ only by scalar "factors", which does not affect the construction of infinite extremals. Therefore it is this case for which the overtaking and the Bellman definitions of infinite extremals agree with each other. Later we will show that Bellman operators with strictly concave kernels (the case specific to economics) actually possess a unique "eigenfunction". This is the reason why the concave case was most extensively investigated.

§9. Concave kernels and dissipative semigroups

Now we turn to the most investigated case of discrete time optimization problems with concave and strictly concave transition functions. For the latter case there exists a complete description both of infinite extremals and of the dynamics generated by the associated Bellman semigroup in the functional space. This dynamics turns out to be dissipative: this means that after normalization any orbit $B^t f$ converges to a unique "eigenfunction" of the operator.

9.1. The framework. From now on we will assume that:

(1) the phase space X is a convex compact subset of a finite-dimensional Euclidean space $\mathbb{R}^n$;

(2) the transition function $b(\cdot\,,\cdot)$ is continuous and strictly concave on $X \times X$ (the strict concavity of a function f on a set X means that for any $x, y \in X$ and for any α, $0 \leqslant \alpha \leqslant 1$, we have

$$f(\alpha x + (1-\alpha y)) \geqslant \alpha f(x) + (1-\alpha) f(y),$$

the strict inequality holding for all $\alpha \neq 0, 1$);

(3) the restriction on the diagonal $b(x, x)$ attains its maximum in a certain *interior* point x^* of X.

From the strict concavity it immediately follows that the point defined by the third condition is unique.

9.2. The equilibrium prices and strictly nonpositive kernels.

LEMMA 9.1. *Under the above set of assumptions there exists a linear functional* $p \in \mathbb{R}^{n*}$, *called the equilibrium price system (for certain economic reasons), such that*

$$\forall x, y \in X, \quad b(x, y) \leqslant b(x^*, x^*) + \langle p, y - x\rangle \tag{9.1}$$

(here $\langle \cdot\,, \cdot \rangle$ *stands for the usual scalar product in* $\mathbb{R}^n$ *).*

PROOF. This fact is a consequence of the standard separation theorem in $\mathbb{R}^n$: the (convex) interior of the hypograph of the function b in $\mathbb{R}^n \times \mathbb{R}^n \times \mathbb{R}$ does not intersect the subspace $\{x = y,\ z = b(x^*, x^*)\}$, hence there exists a linear hyperplane which also does not intersect the interior of the hypograph and contains the above subspace. Due to the interiority condition for x^*, this hyperplane is not vertical, therefore it has an equation of the form $z = b(x^*, x^*) + \langle p, y - x\rangle$ for a certain linear functional p. □

COROLLARY 9.1. *In the assumptions of Lemma 9.1, the associated Bellman operator is equivalent to an operator with strictly nonpositive kernel*

$$\forall x, y \in X,\ b(x, y) \leqslant 0, \quad b(x, y) = 0 \iff x = y = x^*. \tag{9.2}$$

PROOF. Indeed, let us carry out the transformation

$$b \mapsto b(x, y) - \langle p, y\rangle + \langle p, x\rangle - b(x^*, x^*).$$

The last condition holds, since otherwise we would have a segment along which the transition is exactly linear; this contradicts the strict concavity. □

COROLLARY 9.2. *The kernel* (9.2) *actually satisfies the stricter condition*:

$$\forall \varepsilon > 0,\ \exists \delta > 0:\ \operatorname{dist}(x, x^*) \geqslant \delta \implies \forall y \in X,\ b(x, y) \leqslant -\varepsilon. \tag{9.3}$$

PROOF. This comes from the continuity of the kernel (9.2). □

COROLLARY 9.3. *If the optimization problem possesses a transition function satisfying the main assumptions, then any good trajectory* $\mathbf{x}$ *converges to the point* x^* :

$$\lim_{t\to\infty} x_t = x^*. \tag{9.4}$$

PROOF. Indeed, otherwise by Corollary 9.2 the sum $\sum_{t=0}^{\infty} b(x_t, x_{t+1})$ taken over $\mathbf{x}$ diverges to $-\infty$. □

9.3. Overtaking trajectories in the strictly concave case. Since in the strict concave case all the good trajectories converge to the same limit (and the other trajectories are not worth considering), we have an optimization problem with almost fixed endpoints: the condition (9.4) means that the right endpoint is permanently fixed "at infinity". Since such problems are invariant under conjugations (see §2), one can justify the transition to a conjugate operator (9.2).

THEOREM 9.1. *Let* B *be a Bellman operator satisfying the main assumptions. Suppose that there are two good infinite trajectories* $\mathbf{x}$, $\mathbf{y}$ *such that one of them overtakes the other* $\mathbf{x} \overset{B}{\geqslant} \mathbf{y}$ *(in particular this means that they have the same initial point). If* C *is any invertible operator, then the overtaking property is invariant under conjugation by* C :

$$T_C(\mathbf{x}) \overset{C^{-1}\circ B\circ C}{\geqslant} T_C(\mathbf{y}).$$

PROOF. Without loss of generality we may think of C as being a purely diagonal operator, therefore the associated cost flows corresponding to B and $C^{-1}\circ B\circ C$ differ by the expressions $f(x_N) - f(x_0)$ and $f(y_N) - f(y_0)$ respectively. Since the function f is continuous, and both x_N and y_N have the common limit x^*, while $x_0 = y_0$, this modification does not change the limit condition (8.1) from Definition 8.1. □

COROLLARY 9.4. *If* B *is a Bellman operator satisfying the strict concavity assumptions, then there always exist overtaking and weakly optimal trajectories for the corresponding optimization problem, starting at any point of the phase space.*

PROOF. Theorem 9.1 implies that when speaking about good trajectories, without loss of generality one can replace the initial problem by any equivalent one. Therefore we can assume that the transition function is good nonpositive.

Take any infinite trajectory $\mathbf{x}$ with the identically zero cost flow. Since the cost flow of any other trajectory is nonpositive, it is evident that:

(1) $\mathbf{x}$ overtakes any other trajectory;
(2) $\mathbf{x}$ is not supertaken by any other trajectory.

Therefore, returning to the initial transition function, we conclude that $\mathbf{x}$ overtakes and is not supertaken by any good trajectory. On the other hand, trajectories which are not good cannot supertake the good ones and are always overtaken by them. □

9.4. Dissipative dynamical semigroup. Up to now we have discussed the properties of infinite extremals in discrete time optimization problems with strictly concave transition functions. But this concavity implies certain very characteristic properties of the dinamical system on $C(X)$ generated by the corresponding Bellman operator

$$f \mapsto Bf. \tag{9.5}$$

As usual, we will assume that the operator B is normalized so that $\operatorname{spec}(B) = 0$.

THEOREM 9.2. *Suppose that a compact normalized Bellman operator B satisfies the strict concavity assumptions. Then:*

(1) *there exists a unique continuous concave "eigenfunction" $\widehat{f}$;*
(2) *any orbit $B^t f$ converges to $\widehat{f}$ in the uniform topology:*

$$\forall f \in C(X), \quad \lim_{t\to\infty} \|B^t f - \widehat{f}\|_{C(X)} = 0;$$

(3) *the iterates B^t converge to a "rank 1" operator $\widehat{B}$; the kernel of this limit operator can be represented as $\widehat{b}(x, y) = \widehat{f}(x) \odot \widehat{g}(y)$.*

PROOF. 1. Consider the two sequences of functions

$$f_N(x) = \max_{\mathbf{x}:\, x_0=x,\, x_N=x^*} \sum_{t=0}^{N-1} b(x_t, x_{t+1}),$$

$$g_N(y) = \max_{\mathbf{x}:\, x_0=x^*,\, x_N=y} \sum_{t=0}^{N-1} b(x_t, x_{t+1}).$$

Clearly, one has the following monotonicity:

$$f_{N+1}(x) \geqslant f_N(x), \qquad g_{N+1}(x) \geqslant g_N(x).$$

Indeed, any trajectory yielding a maximum in the definition of f_N can be augmented by the final step (x^*, x^*) which adds zero to the value of the criterion, leaving the $(N+1)$-step path admissible with respect to the boundary conditions. The same procedure can be applied to the other sequence (augmentation by the trivial initial step).

On the other hand, both sequences are bounded from above; hence there are uniform limits $\widehat{f}$, $\widehat{g}$. All the functions (including the limit ones) are concave and even strictly concave on X.

Finally, one has the identities $Bf_N = f_{N+1}$, $B^* g_N = g_{N+1}$, following directly from the definition. Passing to limit, one obtains two "eigenfunctions" $B\widehat{f} = \widehat{f}$, $B^*\widehat{g} = \widehat{g}$.

2. Any trajectory solving the finite horizon problem

$$\sum_{t=0}^{N-1} b(x_t, x_{t+1}) + f(x_N) \to \max, \qquad x_0 = x \in X, \tag{9.6}$$

has to come sufficiently close to the point x^*, provided that the time horizon N is large enough:

$$\forall \delta > 0, \ \exists N \in \mathbb{N} : \ \mathbf{x} \in \operatorname{extr}_N(B, f) \implies \exists t, \ 0 \leqslant t \leqslant N : \|x_t - x^*\| \leqslant \delta.$$

Indeed, the value of the problem (9.5) is as close to

$$\widehat{f}(x) + \max_{y \in X}(\widehat{g}(y) + f(y)) \tag{9.7}$$

as we wish, provided that N is large enough. On the other hand, if the entire trajectory $\mathbf{x}$ is outside the δ-ball centered at x^*, then the value of (9.5) does not exceed $-N\delta + \min_{y \in X} f(y)$ and for N large this is a contradiction with (9.7), which gives an estimate independent of N. Moreover, for any given δ the above reasoning proves that only a bounded number of points of the trajectory $\mathbf{x}$ can be outside the δ-ball, with the bound depending on δ, f and uniform over all the initial points $x \in X$.

3. Therefore there must be a moment t such that both t and $N - t$ are large and x_t is close to x^*. Hence the value of the first segment is fairly approximated (for N large) by $\widehat{f}(x)$, the value of the second segment is close to $\widehat{g}(x_N)$, and, since the right endpoint is free, the value of the entire problem (9.5) tends to

$$\widehat{f}(x) + \max_{y \in X}\{\widehat{g}(y) + f(y)\}.$$

4. The uniqueness of the "eigenfunctions" $\widehat{f}$, $\widehat{g}$ follows from the limit assertion. □

Problem. In general, the uniqueness itself cannot guarantee the dissipative behavior of the system. Is it true that the convergence of all the iterations B^t to a certain "rank 1" limit implies the existence of overtaking extremals?

§10. Continuous time case

The theory established above for discrete time optimization problems can be reformulated to cover the continuous time case. In particular, we are interested in problems of the form

$$\int_0^T L(x(t), \dot{x}(t))\, dt + f(x(T)) \to \max, \tag{10.1}$$
$$x(0) = a, \ \ \infty > T \in \mathbb{R}, \ \ x(\cdot) \text{ is absolutely continuous on } [0, T].$$

In order to carry out such a reformulation, we have to assume certain regularity of the integrand $L(x, v)$ independent of t. We assume that the global maximum in the problem (10.1) is attained on an absolutely continuous arc for any initial condition a and any finite time horizon $T < \infty$ (explicit conditions on L guaranteeing such a behavior can be formulated, but we are not interested in them).

DEFINITION 10.1. For any T the *Bellman operator* $B^T : C(X) \to C(X)$ is defined as follows:

$$(B^T f)(a) = \max_{x(\cdot)} \int_0^T L(x(t), \dot{x}(t))\, dt + f(x(T))$$

provided that the maximum is attained.

Clearly, B^T is indeed the Bellman operator with kernel

$$b^T(x, y) = \max_{x(\cdot)} \left\{ \int_0^T L(x(t), \dot{x}(t))\, dt : x(0) = x,\ x(T) = y \right\}.$$

Since the Lagrange function L is independent of time, we have the following fundamental property:

$$B^T \circ B^S = B^{T+S}, \tag{10.2}$$

which means that the family of operators B^T constitutes a one-parameter semigroup, if one puts $B^0 = \mathrm{id}$ by definition.

DEFINITION 10.2. A function $f \in C(X)$ is called an "*eigenfunction*" for a Bellman semigroup B^t, $t \geqslant 0$, if

$$\forall t \geqslant 0, \quad B^t f = f + t\lambda \tag{10.3}$$

for a certain $\lambda \in \mathbb{R}$ which is called the "eigenvalue" of the semigroup.

The difficulty of the continuous time case, which distinguishes it from the discrete time one, is the absence of the generating element. Nevertheless the main result claiming existence of the "eigenfunction" in the special case still holds.

DEFINITION 10.3. A family B^t of Bellman operators is called the *continuous semigroup*, if

(1) $\forall t > 0$, $B^t \in \operatorname{End} C(X)$ is a compact Bellman operator;
(2) $B^t \circ B^s = B^s \circ B^t = B^{t+s}$;
(3) $\lim_{t \to 0^+} \|B^t f - f\| = 0$ for any $f \in C(X)$.

THEOREM 10.1. *A continuous semigroup B^t always possesses a continuous "eigenfunction".*

PROOF. Consider the sequence $t_k = 1/2^k$ and the corresponding operators $B_k = B^{t_k}$. Each of them is compact by definition, therefore it possesses an "eigenfunction" $f_k \in C(X)$: $B_k f_k = f_k + \lambda_k$, $k = 0, 1, \ldots$. Since $B_k \circ B_k = B_{k-1}$, we see that:

(1) each f_k is an "eigenfunction" for all B_j, $0 \leqslant j \leqslant k$;
(2) all the f_k belong to a uniformly continuous subset in $C(X)$ (actually, the latter is the image of B_0);
(3) $2\lambda_k = \lambda_{k-1}$.

Hence there exists a converging subsequence of the form $\widehat{f}_{k_i} = f_{k_i} + c_i \to \widehat{f}_\infty$. By construction, for infinitely many values of k one has $B_k \widehat{f}_\infty = t_k \lambda + \widehat{f}_\infty$, where $\lambda = \lambda_0$. Therefore for all binary rational values of t we have $B^t \widehat{f}_\infty = t\lambda + \widehat{f}_\infty$. By continuity of the semigroup the latter identity holds also for all nonnegative real t. □

In the same manner as this was done in the discrete time case, Theorem 10.1 provides a background for developing the theory of continuous infinite time extremals, parallel to that for the discrete time case.

The continuity assumptions can be verified in several cases, the most important one being the case of a Lagrange function that is strictly concave in all its variables and has the compact domain

$$\operatorname{dom} L = \{(x, v) \in \mathbb{R}^n \times \mathbb{R}^n : L(x, v) > -\infty\}.$$

For details see [29, 32, 17].

Remark. An interesting question (in fact, a whole series of them) arises when analyzing connections between the abstract definition of a semigroup and the particular case of semigroups generated by integral functionals of the form (10.1). There are examples of continuous semigroups which *are not generated* by any function L. The second group of problems concerns differentiability of the Bellman semigroup B^t in the time variable t: it is likely that just as the ordinary linear semigroups of operators are differentiable almost everywhere, so are the Bellman semigroups (at least, under some reasonable assumptions). The last (but not least) problem concerns the possibility to embed a Bellman operator in a continuous semigroup: for a given $B \in \operatorname{End} C(X)$ find a semigroup B^t so that $B^1 = B$.

Some of these questions have been partially answered, but these topics go beyond the scope of the present article.

§11. Survey and concluding remarks

11.1. Overview of the reference list. The algebraic structures of semiring and semimodule associated with combinatorial and graph optimization problems were independently discovered many times: some of the sources are listed in [8]. The recent publications [9, 19, 10, 13] on this subject treat these structures in more detail, paying attention to functional-theoretic properties of the "linear" operations $\oplus$, $\odot$.

Together with our standard assumptions $\odot = +$, $\oplus = \max$ there can be other pairs of operations satisfying the axioms of a commutative distributive semiring. The most important of them is the pair $\oplus = \max$, $\odot = \min$ (see [9, 19]). These operations are used in network flow theory.

Our main technical result, the existence of "eigenfunctions", also has a long history. In 1967 it was proved by I. Romanovskiĭ [16] for the case when the phase space X is a discrete compact set (the "matrix" case) using combinatorial arguments and the duality theory for the linear programming problem

$$\lambda \to \inf, \quad b_{ij} + f_j \leqslant f_i + \lambda, \; i, j = 1, \ldots, n, \; -\infty \leqslant f_j < +\infty, \; \lambda \in \mathbb{R}.$$

In the same paper the notion of maximal loop was introduced and the structure of the "eigenspace" was analyzed.

Another paper by Romanovskiĭ [11] deals with the general case of compact phase space X. He proved the asymptotic formula for iterates

$$B^t f = t\lambda + O(1), \qquad t \in \mathbb{N}, \; t \to \infty,$$

using only an approximation technique. The paper also reveals connections of the spectral problem to the problem of mass transfer.

The most general case of the spectral problem was investigated in papers by P. N. Dudnikov, S. N. Samborskiĭ e.a. [33, 20]. They analyze the axiomatics of idempotent semimodules and formulate conditions, imposed on the pair of operations, sufficient for discrete endomorphisms to have "eigenfunctions". The nondiscrete compact case was investigated by using nonstandard analysis.

The contents of §§5, 6, 8 is new (as far as we know). The idea of cyclic optimization goes back to Romanovskiĭ, and the "trace" formula can also be found in [11]. Actually, this paper, abundant with ideas, is rather difficult to understand.

The idea to use the Bellman principle to generate infinite extremals is rather common: see [21], where the sum of an infinite series was used for this purpose. Other sources: [4, 22]. The idea of transforming the transition function by adding a total difference (total differential to the integrand) in order to describe infinite extremals was exploited in [5, 6].

Another way to approximate the Bellman operator by strict contractions

$$B_\delta f = \max_{y \in X} b(x, y) + \delta f(y), \qquad 0 < \delta < 1,$$

was suggested in [23]. This approach corresponds to investigation of the well-posed problem

$$\sum_{t=0}^{\infty} \delta^t b(x_t, x_{t+1}) \to \max, \qquad x_0 = \text{fixed}.$$

The literature devoted to optimization problems with concave kernels is immense: without going into details we mention here the articles [2, 3, 7, 24–26]. In [17] the nonuniqueness of the "eigenfunction" in the concave but not strictly concave case, arising in mathematical economics, is discussed.

11.2. Miscellaneous topics. Concluding the paper, we mention some relevant results. The detailed exposition will appear elsewhere.

11.3. The homogeneous case: von Neumann-Gale models. One of the very popular models in mathematical economics is reduced to the following optimization problem. Let $K \subset \mathbb{R}^n_+ \times \mathbb{R}^n_+$ be a closed convex cone containing no elements of the form $(0, x)$, $x \in \mathbb{R}^n_+ \setminus \{0\}$. Fix any initial state $x_0 \in \mathbb{R}^n_+$ and any terminal criterion $p \in \mathbb{R}^{n*}_+$ and consider the problem

$$\langle p, x_N \rangle \to \max, \qquad (x_t, x_{t+1}) \in K, \ t = 0, \dots, N-1, \qquad x_0 = \text{fixed}.$$

This finite horizon problem is well posed, but the question is how one can define infinite horizon extremals in this case. There is a definition based on the Pareto partial order on $\mathbb{R}^n_+$, and the existence theorem is proved under some additional assumptions about the cone K.

It is possible to reduce the problem to the standard Bellman framework by letting the phase space X be the standard simplex $\sigma = \{x \in \mathbb{R}^n_+ : x_1 + \cdots + x_n = 1\}$ and introducing the transition function

$$b(x, y) = \max_{\lambda \in \mathbb{R}^1_+} \{\lambda : (x, \lambda y) \in K\}, \qquad x, y \in \sigma.$$

It turns out that under reasonable assumptions this system is dissipative in our sense, thus there exists a natural way to define infinite extremals (or, more precisely, their projections on the simplex), and this definition coincides with the classical one [34].

The analysis is based on ideas suggested in [18].

11.4. Quasicompact case. Most of the theory developed above for the compact Bellman operators works almost without modifications for operators which themselves are not compact, but which have compact powers.

11.5. Game-theoretic interpretation of the projector Ω. The projector Ω: $C(X) \to \mathrm{es}(B)$ introduced in §6 and the corresponding infinite extremals admit interpretation in terms of game theory. Without going into details, we shall say only that an infinite trajectory $\mathbf{x} \in \mathrm{extr}_\infty(\Omega f, B)$ is the path which must be followed in solving the optimization problem

$$\sum_{t=0}^{N-1} b(x_t, x_{t+1}) + f(x_N) \to \max, \qquad x_0 = \text{fixed}$$

with the a priori unknown time horizon N.

11.6. Continuous time semigroups. The concluding remark in §10 lists some problems arising in the theory of continuous semigroups. We can add here that continuous semigroups on a discrete compact phase space can be easily described: the associate matrices must have the form

$$\|b^t_{ij}\| = \begin{pmatrix} t\lambda_1 & 0 & \dots & 0 \\ 0 & t\lambda_2 & \dots & 0 \\ \vdots & \vdots & \ddots & \vdots \\ 0 & 0 & \dots & t\lambda_n \end{pmatrix},$$

while all the other elements are equal to $-\infty$.

11.7. Hamilton-Jacobi theory. In the continuous time case for the variational problem (10.1) there is a relationship between the "eigenfunctions" of the Bellman semigroup and Lagrangian invariant manifolds for the associated Hamiltonian system

$$\dot{x} = \frac{\partial H}{\partial p}, \qquad \dot{p} = -\frac{\partial H}{\partial x}$$

with $H(x, p) = \max_{v \in \mathbb{R}^n} \{\langle p, v\rangle + L(x, v)\}$. The simplest case admitting a complete description is the case of a concave function L. If the Hamiltonian H turns out to be smooth, then the function $f \in C^2(X)$ is the "eigenfunction" for the Bellman semigroup if and only if its gradient graph

$$\{(p, x) \in \mathbb{R}^n \times \mathbb{R}^{n*} : p = \partial f/\partial x\}$$

is an n-dimensional invariant manifold for the Hamiltonian system that passes through a saddle-type singular point of it. For more details see [17].

References

1. L. E. Stern, *Criteria of optimality in the infinite-time optimal control problem*, J. Optim. Theory Appl. **44** (1984), no. 3, 497–508.

2. W. A. Brock and A. Haurie, *On existence of overtaking optimal trajectories over an infinite time horizon*, Mathematics Oper. Res. **1** (1976), 337–346.

3. L. McKenzie, *Turnpike theory*, Econometrica **44** (1976), no. 5, 841–866.

4. V. Z. Belen′kiĭ, *Objective functionals in stationary economic dynamic models*, Modelling of exchange mechanisms in economic systems, Research Report, Central Economics and Mathematics Institute, Moscow, 1979, pp. 93–121. (Russian)

5. D. A. Carlson, *A Carathéodori-Hamilton-Jacobi theory for infinite horizon optimal control problems*, J. Optim. Theory Appl. **48** (1986), no. 2, 93–121.

6. ———, *On the existence of catching up optimal solutions for Lagrange problems defined on unbounded intervals*, J. Optim. Theory Appl. **49** (1986), 207–225.

7. A. Haurie, *Existence and global asymptotic stability of optimal trajectories for a class of infinite horizon non-convex systems*, J. Optim. Theory Appl. **31** (1980), no. 4, 515–533.

8. M. Gondran, *Path algebra and algorithms*, Combinatorial Programming: Methods and Applications (B. Roy, ed.), NATO Adv. Study Inst. Ser., Ser. C: Math. and Phys. Sci., vol. 19, Reidel, Dordrecht, 1975, pp. 137–148.

9. V. P. Maslov, *Asymptotic methods for solving pseudodifferential equations*, "Nauka", Moscow, 1987. (Russian)

10. V. N. Kolokoltsov and V. P. Maslov, *The general form of endomorphisms in the space of continuous functions with values in a numerical semiring with idempotent addition*, Dokl. Akad. Nauk SSSR **295** (1987), no. 2, 283–287; English transl., Soviet Math. Dokl. **36** (1988), no. 1, 55–59.

11. I. V. Romanovskiĭ, *Asymptotic behavior of a discrete deterministic process with continuous state space*, Optimal. Planirovaniye **8** (1967), "Nauka", Novosibirsk, 171–193. (Russian)

12. M. A. Krasnosel′skiĭ, E. A. Lifshits, and A. V. Sobolev, *Positive linear systems*, "Nauka", Moscow, 1985; English transl., Heldermann Verlag, Berlin, 1989.

13. V. P. Maslov, *On a new superposition principle for optimization problems*, Uspekhi Mat. Nauk **42** (1987), no. 3, 39–48; English transl. in Russian Math. Surveys **42** (1987).

14. M. G. Crandall and P. L. Lions, *Viscosity solution of Hamilton-Jacobi equations*, Trans. Amer. Math. Soc. **277** (1983), 1–42.

15. Z. Nitecki, *Differentiable dynamics*, MIT Press, Cambridge, MA, and London, England, 1971.

16. I. V. Romanovskiĭ, *Optimization of stationary control of discrete deterministic processes in dynamic programming*, Kibernetika (Kiev) **2** (1967), 66-78; English transl. in Cybernetics **3** (1967).

17. S. Yu. Yakovenko, *Fixed points of a semigroup of Bellman operators, and invariant manifolds of Hamiltonian subdifferential equations*, Avtomat. i Telemekh. **1989**, no. 6, 43–52; English transl. in Automat. Remote Control **50** (1989).

18. A. D. Tsvirkun and S. Yu. Yakovenko, *Lyapunov functions and turnpike theory*, Lecture Notes in Control and Inform. Sci., vol. 84, Springer-Verlag, Berlin and New York, 1986, pp. 1002–1007.

19. S. M. Avdoshin *et al.*, *Optimization of flexible production systems*, Preprint, Moscow Institute of Electronic Engineering, Moscow, 1987. (Russian)

20. P. I. Dudnikov and S. N. Samborskiĭ, *Endomorphisms of semimodules over semirings with an idempotent operation*, Preprint No. 87-48, Inst. Mat. Akad. Nauk Ukrain., Kiev, 1987. (Russian)

21. M. Weitzman, *Duality theory for infinite horizon convex models*, Management Sci. **19** (1973), no. 7, 783–789.

22. F. H. Clarke and Vinter R. B., *The relationship between the maximum principle and dynamic programming*, SIAM J. Control Optim. **25** (1987), no. 5, 1291–1311.

23. E. Denardo, *Contraction mapping in the theory underlying dynamic programming*, SIAM Rev. **9** (1967), 165–177.

24. L. McKenzie, *Turnpike theorems with technology and welfare functions variable*, Mathematical Models in Economics, Academic Press, New York and Amsterdam, 1974, pp. 271–278.

25. A. Haurie, *Optimal control on an infinite time horizon: the turnpike approach*, J. Math. Econom. **3** (1976), no. 1, 81–102.

26. B. Peleg and H. Ryder, *On optimal consumption plans in a multisector economy*, Rev. Econom. Stud. **39** (1972), no. 2, 159–169.

27. R. T. Rockafellar, *Saddle points of Hamiltonian systems in convex problems of Lagrange*, J. Optim. Theory Appl. **12** (1973), no. 4, 367–390.

28. N. N. Vorob′ev, *Extremal algebra of nonnegative matrices*, Elektronische Informationsverarbeitung und Kybernetik **6** (1970), 303–312.

29. A. Leizarowitz, *Infinite horizon autonomous systems with unbounded cost*, Appl. Math. Optim. **13** (1985), 19–43.

30. D. Gale, *On optimal development in a multi-sector economy*, Rev. Econom. Stud. **34** (1967), 1–19.

31. C. C. von Weizsacker, *Existence of optimal programs of accumulation for an infinite horizon*, Rev. Econom. Stud. **32** (1965), 85–104.

32. A. Leizarowitz, *Existence of overtaking optimal trajectories for problems with convex integrands*, Math. Oper. Res. **10** (1985), no. 3, 450–461.

33. P. I. Dudnikov and S. N. Samborskiĭ, *Spectra of endomorphisms of semimodules over semirings with an idempotent operation*, Dokl. Akad. Nauk SSSR **308** (1989), 780–783; English transl. in Soviet Math. Dokl. **40** (1990).

34. A. M. Rubinov, *Economic dynamics*, J. Soviet Math. **26** (1984), no. 4.

INSTITUTE OF CONTROL SCIENCES, MOSCOW, RUSSIA

ELSINT, HAIFA, ISRAEL

Translated by S. YU. YAKOVENKO